AF597992

MICROELECTRONICS

Micro and Nano-electronics and Photonics

MICROELECTRONICS
Micro and Nano-electronics and Photonics

— Editor —
Krishan Lal

CENTRE FOR SCIENCE & TECHNOLOGY OF THE NON-ALIGNED AND OTHER DEVELOPING COUNTRIES
(NAM S&T CENTRE)

2009
DAYA PUBLISHING HOUSE
Delhi - 110 035

ISBN 81-7035-622-9
ISBN 978-81-7035-622-6

Centre for Science and Technology of the Non-Aligned and Other Developing Countries (NAM S&T Centre)
Core-6A, 2nd Floor, India Habitat Centre, Lodhi Road,
New Delhi-110 003 (India)
Phone: +91-11-24644974, 24645134, Fax: +91-11-24644973
E-mail: namstct@gmail.com
Website: www.namstct.org

Published by : **Daya Publishing House**
1123/74, Deva Ram Park, Tri Nagar, Delhi - 110 035
Phone: 27383999, Fax: (011) 23260116
e-mail: dayabooks@vsnl.com
website: www.dayabooks.com
Showroom : 4760-61/23, Ansari Road, Darya Ganj
New Delhi - 110 002
Phone: 23245578, 23244987

Laser Typesetting : **Classic Computer Services**
Delhi - 110 035

Printed at : **Chawla Offset Printers**
Delhi - 110 052

Printed in India

Foreword

There have been revolutionary technological changes over the past six decades that have transformed our way of life, largely with the shrinking of distance and the speed at which everything can be accomplished. We are now moving into a new dimension of space and time: that which characterizes cyber space. Our capabilities to access the knowledge base of the world has been with us for several thousand years since the invention of writing.

However, it was in 1948, that there was an announcement from the Bell Laboratories of the discovery of the transistor by Bardeen, Brattain and Shockley. This was a small solid state device that would replace the bulky vacuum tube technology which consumed a lot of energy and gave out a lot of heat. However, there was still with us what was referred to as a 'tyranny of numbers', which meant that a large number of separate components like transistors, resistors, capacitators, and the like would have to be connected mechanically.

And then, there was the remarkable discovery during 1958-59 of what is referred to as the integrated circuit. This was the work of two wholly independent scientists: one was Jack Kilby who was working in Texas Instruments and demonstrated that the various circuit components could be manufactured on a single piece of semi-conductor material. Jack S. Kilby was awarded the Nobel Prize for 2000 for this invention; his patent for a solid state circuit made of germanium was filed in February 1959. There was, in parallel, the work of Robert Noyce, who had first worked with William Shockley (discoverer of the transistor) at the Shockley Semi-conductor Laboratory; he then went on to create Fairchild Semi-conductors. In 1959 Robert Noyce was awarded his patent for the discovery of the integrated circuit. Since Noyce died in 1990, he could not share the Nobel Prize with Jack S. Kilby which, otherwise, he would have done.

Since then, there have been continuing spectacular advances in this new field of micro-electronics, that have taken us through small scale to medium scale, large scale and ultra large scale integration. Every 18 months, chip performance has gone up by a factor of 2, and there has been an equal reduction in cost by a factor of 2. It is estimated that in a decade, at the same cost, processing capabilities would go up by a factor of 100, storage by a factor of 1000 and band width by a factor of 10,000. This continuing advance is awesome; completely transforming all areas of human activities: business, finance, industry, science and technology and, indeed, one's personal life.

Since then, great emphasis has continued on improvements in design techniques, in starting materials like silicon, and fabrication steps including lithography, impurity incorporation, interconnects, precision step movements of devices and packaging. Advances in all these areas has enabled continuous decrease in the feature sizes and increase in the number of devices per circuit. Presently, the number of devices per circuit has surpassed one billion, and the feature sizes have moved into the nano-metric scale. These developments bring with them many challenges, in maintaining high quality of starting materials, strict control during several hundred processing steps and design of automated equipment to carry out these steps faultlessly. Therefore, developments in materials synthesis and preparation, characterization and diagnostic techniques and fabrication technology are of utmost importance. Since the rate of growth of this field is very fast, it is necessary to review the status periodically. Micro-electronics S&T is now moving into nano-electronics range, and the subject of photonics has also made an enormous impact, and utilizes several basic technologies of micro-electronics. Both for R&D as well as for production, the infrastructure required is very expensive. Also, highly trained manpower is a necessary requirement. In view of these factors, only a few advanced countries have commercial production in this field.

However, in view of the impact of this technology in all walks of human life, the leadership in the developing world is aware of its importance and wishes to initiate activities, be it in R&D or in production facilities. The level of development in different developing countries is different, and there is a lot to be gained by south-south collaboration. I am glad to know that NAM S&T Centre has been organizing international workshops and training courses in this important area.

In this series, the Sixth International Workshop and Training Course was organized in Islamabad in 2007, in which a large number of experts from advanced countries and scientists from developing world took active part. I compliment the NAM S&T Centre, particularly Dr. A.P. Kulshreshta, its Director for this initiative and also for bringing out the proceedings of this meeting.

This volume has both review papers as well as country reports. Review papers cover important devices like gallium nitride and silicon carbide based high frequency, high power amplifiers, MEMS and advanced materials preparation covering nano materials and silicon carbide films. One paper deals with an important ion beam technique employed for implantation of impurities in semi-conducting crystals. Important advances made in non-destructive characterization of single crystals, thin films and devices by high resolution X-ray diffraction techniques have been discussed. Also, scanning probe microscopy for biological imaging has been described. A possible future system in the important area of consumer electronics is the subject matter of an interesting paper. Country reports from Asia and Africa give an interesting picture of different levels of efforts being made in these countries to promote the field of microelectronics.

This volume will be useful for researchers working on several aspects of microelectronics, nanoelectronics and photonics and also, for planners. The proceeding implicitly underlines the possibility of North-South and South-South collaboration.

I compliment Dr. Krishan Lal, who has put in a great effort, as the Editor, and has organized the book in a systematic and useful manner for different types of readers.

Prof. M.G.K. Menon
Advisor, ISRO / Department of Space, Govt. of India

Preface

Electronics science has experienced several exciting revolutions since the discovery of electron. The emergence of semiconductor devices as replacement of glass tubes was the first major landmark followed by success achieved in integration of electronic circuits on the same crystal substrate by planar technology. The quest to reduce device dimensions took us to sub-micron feature sizes. The number of devices per integrated circuit has been increasing following the famous Moore's Law. We are now dealing with more than a billion devices per integrated circuit. This success has been achieved with the help of spectacular advances in the materials preparation technologies notably nearly perfect silicon substrates with extremely controlled chemical composition and structure and expertise in preparation of thin layers of oxides, metals, and other materials. The deliberated addition of impurities by controlled ion implantation and special infrastructure for high quality lithography processes has also played a crucial role. The phenomenal developments in packaging technologies were essential in meeting ever growing demands. The fundamental understanding of materials and processes and strict control on materials characteristics during device fabrication has been possible with the help of advanced diagnostic techniques and equipments. The developments in microelectronics and now nanoelectronics are touching practically all aspects of human life. One cannot imagine practically any modern equipment without microelectronic devices, be these, computers, cell phones, TVs, communication infrastructure or medical equipment. Unfortunately, destructive weapons also extensively employ these devices.

Microelectronics is one of the priority areas identified by the Science and Technology Centre of the Non-Aligned and other Developing Countries (NAM S&T Centre). A series of Workshops have been organized by the Centre in different countries, which have played an important role in human resource development and growth of this important area in the member countries. The 6th International Workshop on Microelectronics was organized by the NAM S&T Centre at Islamabad during 9-13 April 2007 jointly with the Institute of Information Technology (IIT) of the Commission on Science and Technology for Sustainable Development in the South (COMSATS). It was truly an international meeting with scientists from US, Europe, Asia and Africa participating in the deliberations. Several top class experts covered topics of considerable importance in the subject matter of the Conference. In addition, country reports were also presented. The interactions between the lecturers and the young and enthusiastic participants particularly those from Pakistan were very stimulating. The discussions continued even after the formal programme was over. The hosts had made good arrangements, which were conducive for discussion among the scientists.

All the invited speakers and national representatives were requested to make available manuscripts of their presentations. As a result of this effort, ten papers have been received which covered a wide variety of topics including those on devices, processing steps, thin layers and nano materials as well as advanced techniques for characterization related to materials processing and devices. In addition, eight country reports are also presented mainly from Asian and African countries.

Preparation of advance materials has been covered in this proceedings, particularly the tailored thin films and nano materials and growth and characterization of single crystalline cubic SiC on porous Si using low pressure chemical vapour deposition technique. Introduction of desired impurities in semiconductors in highly controlled manner by ion beam technique has been discussed particularly keeping in view the challenges posed by futuristic devices. Advances in gallium nitride and silicon carbide based high frequency power amplifiers and MEMS for understanding mechanical behavior of metal nano films have been discussed. One of the paper deals with system architecture of future consumer electronic systems, which will revolutionize the living styles. A number of papers deal with advanced techniques of characterization like the high resolution X-ray diffraction methods for investigations of single crystals, thin films and devices as well as scanning probe microscopy for biological imaging. Optical pattern recognition and generation by using neural networks has been described in one of the papers.

Interesting country reports from Indonesia, Kenya, Myanmar, Nepal, Sri Lanka, Tanzania, Turkey, and Uganda are included in this Volume. These present the status of microelectronics and related fields in these countries and the efforts being made there to improve human resources, R&D infrastructure and industrial base.

Since the English language used in different countries has variations, I have tried to make editorial changes in several papers and reports, which hopefully improve their readability. Papers on advanced topics will be very useful for active researchers in terms of the results presented there as well as the bibliography. As expected, there are wide variations in the level of development in different countries. However, it is quite clear that all countries are making efforts to enhance the R&D based so as to gain best advantage from this exciting area in their respective countries. Also, it is clear that there is a lot of scope for bilateral and multilateral collaborations. The South-South collaboration needs particular encouragement.

I complement the NAM S&T Centre for organizing this series of Workshops at different locations underlining the importance of this field for fast growth and development of societies in less developed world. It is a pleasure to appreciate the high level of commitment and dynamism with broad understanding exhibited by Dr. A.P. Kulshreshtha, Director, NAM S&T Centre. He was very much involved in different aspects of the organization and encouraging scientists from different countries to develop collaborative activities. Dr. V.P. Kharbanda has been helping in the technical editing of the proceedings and thanks are due to him for his contributions and assistance in editorial work. I am thankful to Smt. Saroj Gandhi for providing able support at NPL throughout the editorial work.

Krishan Lal

Introduction

In the last few decades we have witnessed the realization of many new hitherto unbelievable ideas and concepts that were only found in science fiction. Most of this has been made possible through ultra-micro-miniaturization of the components and systems. Microelectronics is almost a generic word and subsumes areas as diverse as communication, information technology, data processing, process control, nanotechnology, nano-electronics, photonics, etc. to automation of a host of industries as well as biological and medical applications. The advancements in microelectronics fabrication technology have resulted in the development and fabrication of an assortment of high quality and high precision micro-sensors and sensors based on micro-electromechanical systems and devices (MEMS). Microelectronics and Information Technology are the two key elements that have led to the high level of technological sophistication achieved by the present day civilization. Successive generations of miniaturization of electronic systems have resulted not only in the improvement of their performance but also in the economy of scale. Further, low production costs achieved due to the application of batch fabrication technology, small size and low power consumption have facilitated the transformation of the manufacturing of microelectronics and micro-systems into a commercially viable industry. Because of its immense economic impact and commercial potential, almost all the developing countries are gradually aspiring to adopt microelectronics R&D and production. However, at the same time, they are also desperately in need of comprehending the full implications of the level of microelectronics technology presently required by them within the nation and forecasts for the years ahead in terms of its capabilities, limitations, viability, economics and infrastructural requirements by undertaking a pragmatic analysis in totality.

In view of the ever-growing importance of Microelectronics affecting our daily lives, the Governing Council of the NAM S&T Centre has accorded high priority to this area. The Centre has earlier organised 5 international workshops and training programmes on various segments of Microelectronics at almost 1½ -2 years interval.

These were respectively in Bangkok in association with the Asian Institute of Technology, Thailand in August 2000; at Da Nang, Vietnam in November 2001; in Delhi in association with the Central Electronics Engineering Research Institute(CEERI), Pilani and Council of Scientific & Industrial Research (CSIR) of India in December 2002; in Cairo in association with the Electronics Research Institute, Egypt in December 2003; and in Kuala Lumpur in association with MIMOS Berhad of Malaysia in September 2005. Continuing with these endeavours, the NAM S&T Centre had organised the sixth event in the series, that is, the international workshop–cum–training course on Microelectronics with a focal theme on 'Micro- and Nano-electronics and Photonics' during 9-13 April 2007 at Islamabad, Pakistan jointly with the Institute of Information Technology (IIT) of the Commission on Science and Technology for Sustainable Development in the South (COMSATS).

The Islamabad Training Workshop on microelectronics was aimed at facilitating the exchange of country specific information on microelectronics with particular reference to photonics materials and applications, power devices, trends in semiconductor process simulation, system architecture and next generation silicon based electronics, and deliberating on North-South and South-South Cooperation to develop this sector in the developing countries.

Besides the inaugural session, the training workshop was conducted in eight technical, country report and training sessions as well as a day long trip to Ghulam Ishaq Khan Institute of Engineering Sciences and Technology (GIKI), a Center of Excellence in Pakistan for the Natural Sciences and Computing located near Topi to acquaint and train the participants with the state-of-the-art techniques in Microelectronics. Several researchers also displayed posters of their research work during the course of the workshop. The overall technical programme of the training workshop was coordinated by Dr. Ahmed Shuja and Dr. Muhammad Asif Malik of COMSATS-IIT and was attended by about 170 scientists from 14 countries. The overseas participants included one each from Austria, Indonesia, Malaysia, Myanmar, Nepal, Netherlands, Sri Lanka, Turkey, Uganda, UK, two from India and Sweden, and three from USA who made their presentation during the event. Scientists from the host country Pakistan included the Heads and senior experts from COMSATS-IIT, Ghulam Ishaq Khan Institute of Engineering Sciences and Technology (GIKI), CSIR Labs, National Physical and Standards Laboratory and National Institute of Electronics, National Engineering and Scientific Commission (NESCOM), Pakistan Institute of Engineering and Applied Sciences (PIEAS), National University of Science and Technology (NUST), the Islamia University of Bahawalpur, University of Peshawar, University of Agriculture, University of the Punjab, Bahauddin Zakriya University, University of Sargodha, University of Karachi, Allama Iqbal Open University and Quaid-i-Azam University(QAU).

The concluding ceremony was chaired by Dr. Noor Mohammed Butt, Chairman, Pakistan Science Foundation; Prof. Farid A. Khwaja, Director General, National Institute of Electronics, Islamabad; and Prof. Arun P. Kulshreshtha, Director, NAM S&T Centre, before which an Informal Group Discussion moderated by Dr. Ahmed Shuja was held with Dr. Zafar Iqbal, Mrs. Dr. Naseem Zafar, Prof. Kasturi Lal Chopra,

Dr. Krishan Lal, Dr. Noor Mohammed Butt, Prof. Arun P. Kulshreshtha and Prof. Dean Aslam as the panelists.

The present publication comprises a compilation of 18 scientific articles and review papers. I would like to express gratitude to Honourable Dr. Ishfaq Ahmad, Special Advisor to the Prime Minister of Pakistan; His Excellency Prof. Atta-ur-Rahman, the then Federal Minister and Chairman, Higher Education Commission; and Dr. S.M. Junaid Zaidi, Rector of COMSATS-IIT and his colleagues Prof. Arshad Bhatti, Dr. Ahmed Shuja and Dr. Mohd. Asif for their comprehensive support and guidance in this venture. I am also thankful to the Ministry of Science and Technology of Pakistan for holding this event.

Last, but not the least, I would also like to acknowledge the dynamic involvement and untiring effort of Dr. Krishan Lal, formerly, Director, National Physical laboratory of India for technical editing of this publication, supplying the 'Preface' and for suggesting future course of action while reviewing the contents. I am indebted to Prof. M.G.K. Menon, Advisor, ISRO / Department of Space, Govt. of India for sparing his valuable time in writing the 'Foreword'. The valuable services provided by the entire team of the NAM S&T Centre, particularly by Mr. M. Bandyopadhyay, Dr. V. P. Kharbanda, Mr. Pankaj Buttan and Mrs. Sudipta Chakraborty in compiling the presented papers and giving a shape to this volume are also deeply appreciated.

I hope that this publication would be useful for the entire scientific community of the microelectronics researchers, experts and practitioners particularly from the developing countries who wish to develop, implement and intensify programmes and facilities related to microelectronics in their economies through South-South cooperation.

Arun P. Kulshreshtha
Director, NAM S&T Centre

Contents

Chapter 1

GaN and SiC Based High Frequency Power Amplifiers

S. Azam and Q. Wahab*

Department of Physics,
Chemistry and Biology (IFM), Linköping University,
SE-58183 Linköping, Sweden
**E-mail: Sher@ifm.liu.se,quw@ifm.liu.se*

ABSTRACT

The wide band-gap materials have focused the attention of the researchers due to their ability to withstand high temperatures and high frequencies as well as their performances demonstrated for high power applications. The rapid development of the RF power electronics requires the introduction of WBG materials due to extra ordinary properties compare to conventional materials. GaN and SiC-based RF power amplifiers have made substantial progresses in the last decade. This paper attempts to review the latest developments of GaN and SiC Transistors based power amplifier technologies.

Keywords: *Gallium-nitride, HEMTs, Silicon carbide, MESFET, MMICs, Power amplifiers.*

Introduction

Phase array (PA) radars, wireless communication market and other traditional military applications require demanding performance of microwave transistors. In several RF applications, as well as in radar and military systems, the development of circuits and sub-systems with broadband capabilities is always more and more demanding. From transmitter point of view the bottleneck, and consequently the critical key factor, is represented by the development of a high performance PA. The latter, in fact, deeply influences the overall system features in terms of bandwidth, output power, efficiency, working temperature etc. So far, distributed approaches have often been proposed and investigated to design broadband amplifiers also in GaN technology (Gassmann *et al.*, 2007).

Next generation cell phones require wider bandwidth and improved efficiency. The development of satellite communications and TV broadcasting requires amplifiers operating at higher frequencies and higher power to reduce the antenna size of terminal users. The same requirement holds for broadband wireless internet connections as well. These high power and high frequency applications require transistors with large breakdown voltage, high electron velocity and high thermal conductivity. The wide band gap materials, like GaN and SiC are preferable and pre-eminent technology of choice (Mishra *et al.*, 2008).

The output power density of WBG transistors allows the fabrication of much smaller size devices with the same output power. Higher impedance due to the smaller size allows for easier and lower loss matching in amplifiers. The operation at high voltage due to its high breakdown electric field not only reduces the need for voltage conversion, but also provides the potential to obtain high efficiency, which is a critical parameter for amplifiers. The wide band gap also enables it to operate at high temperatures. These attractive features in amplifier applications enabled by the superior semiconductor properties make these devices promising candidates for microwave power applications.

In this article we discuss the power, gain and efficiencies of GaN HEMT and SiC Transistors based power amplifiers.

GaN Based Power Amplifiers

In the last decade, AlGaN/GaN HEMT technology has established itself as a strong contender for the applications (mentioned in Introduction) because of its large electron velocity (>10^7 cm/s), wide band gap (3.4 eV), high breakdown voltage (> 50 V for f_T =50 GHz) and sheet carrier concentration (>10^{13} cm^{-2}). Due to the superior electronic properties of the GaN material and the possibility to use SiC substrate demonstrating high thermal conductivity (4.5 W/cm.K), power densities as high as 30 W/mm @4 GHz (Wu *et al.*, 2004) as well as output power of 500 W @1.5 GHz have already been achieved(Maekawa *et al.*, 2006) .An output power of 75 W of a packaged single-ended GaN-FET has been reported under pulsed conditions for L/S band applications (Wakejima *et al.*, 2006).

The GaN technology is currently widely used for power amplifier applications. For mobile and base station applications a number of manufacturers and researchers have reported high efficiencies, output powers and power densities (Wataru *et al.*, 2006; Kawano *et al.*, 2005; Kim *et al.*, 2008; Moon *et al.*, 2008; Colantonio *et al.*, 2008; Kikkawa *et al.*, 2005; Sheppard *et al.*, 2002; Yamanaka *et al.*, 2005; Kikkawa,2004; Maekawa *et al.*, 2005; Wu *et al.*, 2006; Jyomasa *et al.*, 2007; Otsuka *et al.*, 2004; Ui *et al.*, 2007; Wu *et al.*, 2007; Gassmann *et al.*, 2007; Lee and Jeong2007; Hong *et al.*, 2007; Yamanaka *et al.*, 2007; Tayrani, 2007; Okamoto *et al.*, 2007; Gustavsson *et al.*, 2007; and Lee *et al.*, 2008). A class-E amplifier at 13.56-MHz with a high-voltage GaN HEMT as the main switching device is demonstrated to show the possibility of using GaN HEMTs in high frequency RF power-supply applications. The 380-V/1.9-A

GaN power HEMT was designed and fabricated for high voltage power electronics applications. The circuit demonstrated has achieved the output power of 13.4 W and the power efficiency of 91 per cent under a drain–peak voltage as high as 330 V. This result shows that high-voltage GaN devices are suitable for high-frequency switching applications under high dc input voltages of over 100 V (Wataru *et al.,* 2006).The Eudyna GaN hybrid power amplifier from Eudyna is capable of efficiently delivering 200 W of power at 2.1 GHz for W-CDMA applications (Kawano *et al.,* 2005).A picture and results are shown in Figures 1.1 and 1.2, respectively.

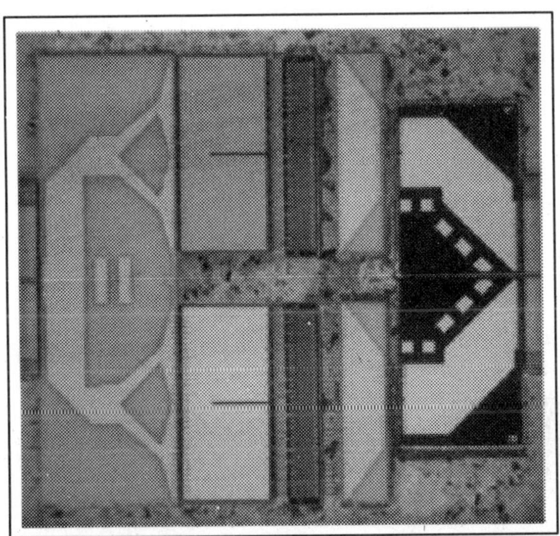

Figure 1.1: Picture of a Large Periphery 2.1 GHz GaN HEMT Hybrid Assembly for W-CDMA Base Station Applications (Eudyna) (Wakejima *et al.,* 2006)

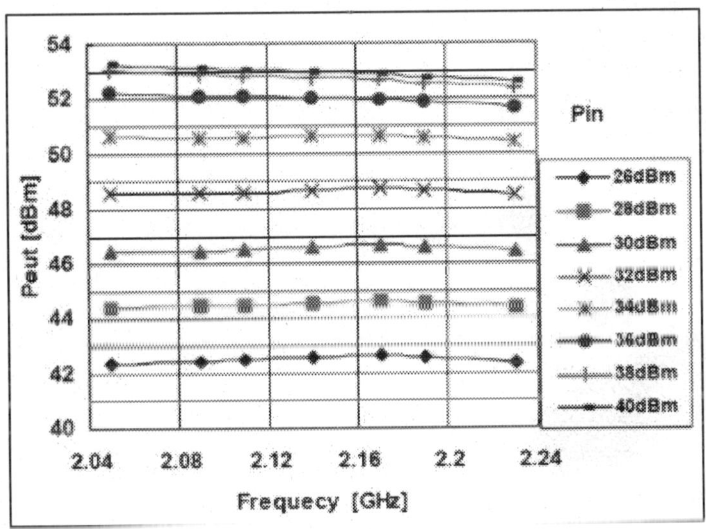

Figure 1.2: Result of Output Power of a Large Periphery 2.1 GHz GaN HEMT Hybrid Assembly for W-CDMA Base Station Applications (Eudyna) (Wakejima *et al.,* 2006)

The saturated Doherty power amplifier implemented using two Eudyna EGN010MK GaN HEMTs with a 10 W peak envelope power for a 2.14 GHz forward-link W-CDMA signal. The amplifier showed an excellent efficiency of 52.4 per cent with an acceptable linearity of 28.3 dBc at an average output power of 36 dBm. Moreover, the amplifier can provide high linearity performance of 50 dBc using the digital feedback pre-distortion technique(Kim *et al.*, 2008). A wideband envelope tracking Doherty amplifier, implemented using Eudyna 10-watt GaN high electron mobility transistor (HEMT) for world interoperability for microwave access (WiMAX) signals of the 802.16d and 802.16e, covering 90 MHz bandwidth. The envelope tracking amplifier delivers a significantly improved relative constellation error (RCE) of 35.3 dB, which is an enhancement of about 4.3 dB, while maintaining the high PAE of 30 per cent for the 802.16d signal with an average output power of 35 dBm (Moon *et al.*, 2008).

An ultra-wide band high efficiency power amplifier (PA) is implemented in GaN technology as shown in Figure 1.3. A HEMT device with 1 mm of gate periphery at 0.8-4 GHz, showing drain efficiency greater than 40 per cent with an output power higher than 32 dBm in the entire bandwidth.

CREE Inc. has also demonstrated compact, high-power microwave amplifiers taking advantages of the high-voltage and high power density of GaN HEMTs

Figure 1.3: Photograph of Power Amplifier (Colantonio *et al.*, 2008)

(Wu *et al.*, 2006). A peak power of 550 W (57.40 dBm) is achieved at 3.45 GHz with 66 per cent DE and 12.5 dB associated gain. An outstanding power-efficiency combination of 521 W and 72.4 per cent is obtained at 3.55 GHz. Such power levels, accompanied by the high efficiencies, are believed to be the highest at around 3.5 GHz for a fully-matched, single-package solid-state power amplifier, attesting the great potential of the GaN technology in high frequency electronics.

The state-of-the-art drain-efficiency over 50 per cent (over 45 per cent power-added-efficiency) of GaN HEMT high power amplifier (Figure 1.4) together with over 50 W output power at C-band is proposed and implemented(Jyomasa *et al.*, 2007) and are as shown in Figure 1.5. A GaN HEMT of 16mm gate periphery will be enough for

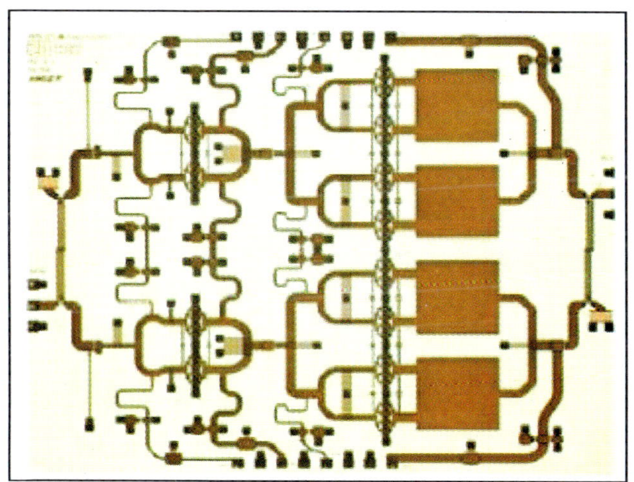

Figure 1.4: Picture of a 31–36 GHz Balanced GaN MMIC Power Amplifier (Jyomasa *et al.*, 2007)

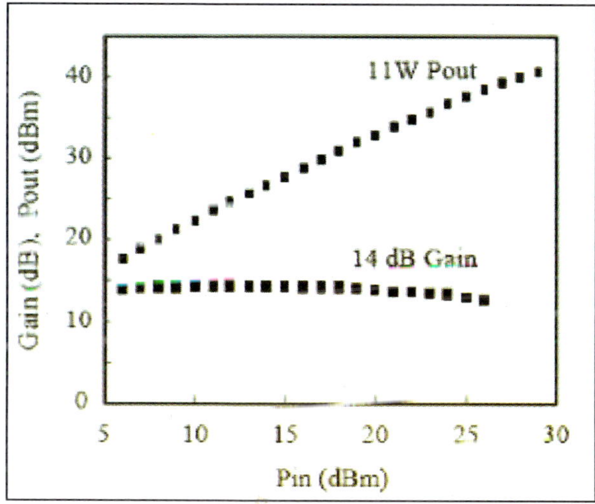

Figure 1.5: Performance of a 31–36 GHz Balanced GaN MMIC Power Amplifier. 11 W of output power at 34 GHz were achieved (Jyomasa *et al.*, 2007).

60 W output power for this power density. Consider that the GaAs technology demands a minimum of 75.6 mm gate-width for only 20 W output power(Otsuka *et al.*, 2004), a 230mm total gate width would be needed to realize 60 W. Therefore, GaN HEMT devices are desirable to realize broadband high power amplifier. A 2-stage amplifier up with the 30 W driver stage, 42 per cent efficiency (including 30W driver amplifier) and–50 dBc ACLR at the average power of 49 dBm (80 W) with saturation power of 56.5 dBm and gain of 32 dB is obtained for WCDMA applications(Ui *et al.*, 2007).A 0.4 mm wide GaN device with 0.15 μm gate and 0.25 μm field plate operated up to 60 V achieved 13.7 W/mm power densities at 30 GHz, the highest for a FET at millimeter-wave frequencies(Wu *et al.*, 2007).

A highly efficient, wide band power amplifier designed in GaN technology and utilizing a non-uniform distributed topology is reported (Gassmann *et al.*, 2007). Measured results demonstrate very high efficiency across the multi-octave bandwidth. Average CW output power and PAE across 2-15 GHz was 5.5 W and 25 per cent, respectively. The MMIC is shown in Figure 1.6. Maximum output power reached 6.9 W with 32 per cent PAE at 7 GHz. (Gassmann *et al.*, 2007). The results obtained for class-E power amplifier using GaN HEMT (Figure 1.7) are; the power added efficiency

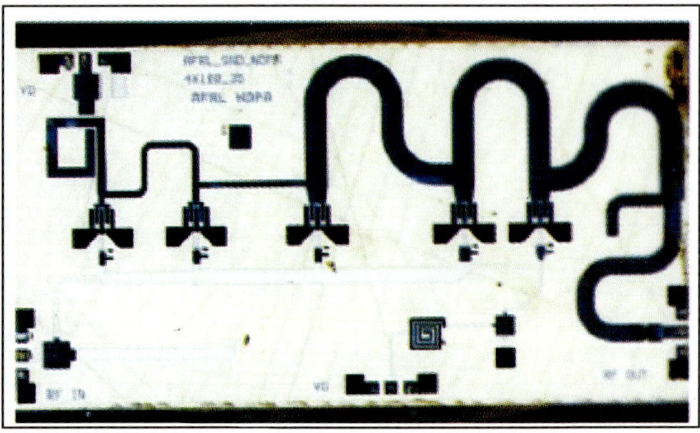

Figure 1.6: Fabricated NDPA MMIC Mounted on a CuW Carrier. Chip size is 4.45 mm x 2.26 mm (10 mm^2) (Gassmann *et al.*, 2007).

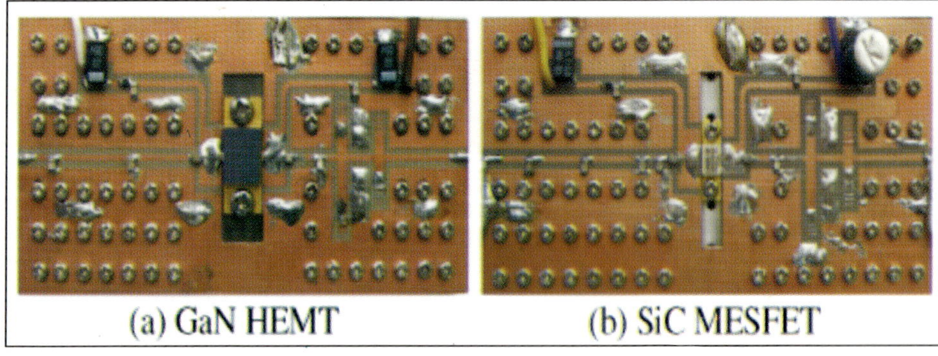

Figure 1.7: Photographs of the Fabricated Class-E Power Amplifier Using Wide-band Gap Devices (Lee and Jeong, 2007)

(PAE) of 70 per cent with a gain of 13.0 dB at an output power of 43.0 dBm, through significant reduction of harmonic power level(Lee and Jeong,2007).

A class F mode PA using Eudyna's GaN HEMT has been biased at class C and adopted a new output matching topology that improved the overall transmitter efficiency. For the WiMAX OFDM signal, the calculated overall drain efficiency of the optimized EER amplifier, based on the measured bias dependent efficiency of 73 per cent at an average power of 31 dBm at 2.14 GHz. The proposed highly efficient bias modulation PA for the EER transmitter provides a superior overall efficiency than that of any conventional switching or saturation mode PAs (Hong *et al.*, 2007).

A 2-chip amplifier has recorded 220 W output power at C-band, which is the highest output power ever reported for GaN HEMT amplifier chip at C-band and higher bands(Yamanaka *et al.*, 2007). A picture and results are shown in Figures 1.8

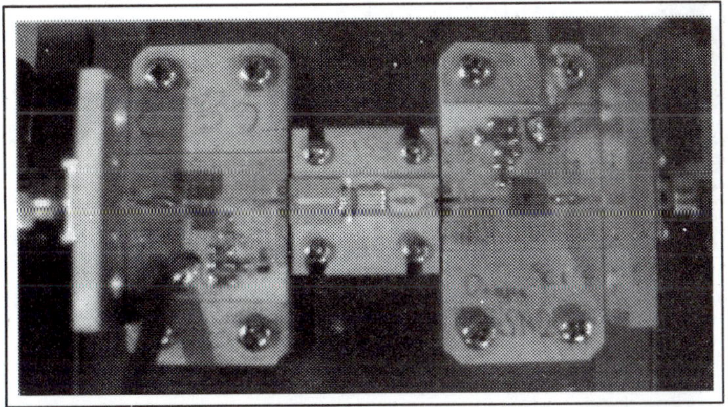

Figure 1.8: Photograph of the C-band Single Chip GaN HEMT Amplifier (Yamanaka *et al.*, 2007)

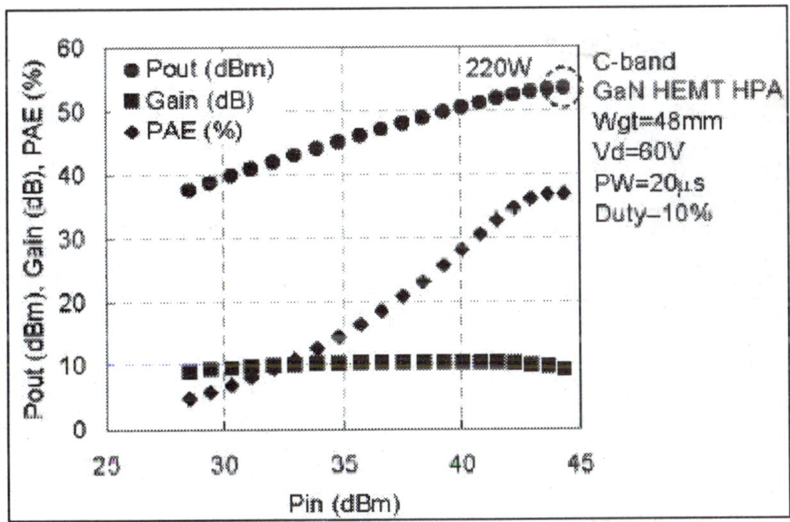

Figure 1.9: Output Characteristics of the C-band 2-chip GaN HEMT Amplifier (Yamanaka *et al.*, 2007)

and 1.9. A highly efficient broadband monolithic class-E power amplifier is implemented utilizing a single 0.25 µm x 800 µm AlGaN/GaN field-plated HEMT producing 8 W/mm of power at 10.0 GHz. The HPA utilizes a novel distributed broadband class-E load topology to maintain a simultaneous high PAE and output Power over (6-12 GHz). The HPA's peak PAE and output power performance measured under three pulsed drain voltages at 7.5 GHz are: (67 per cent, 36.8 dBm @ 20 V), (64 per cent, 37.8 dBm @ 25 V) and (58 per cent, 38.3 dBm @ 30 V) (Tayrani,2007).

A C-band high-power amplifier with two GaN-based FET chips exhibits record output powers under continuous-wave (CW) and pulsed operation conditions (Figure 1.10). At 5.0 GHz, the developed GaN-FET amplifier delivers a CW 208 W output power with 11.9 dB linear gain and 34 per cent power-added efficiency. It also showed a pulsed 232 W output power with 8.3 dB linear gain (Okamoto *et al.*, 2007). A class D-1 amplifier is implemented with GaN MESFETs (Figure 1.11) and working around 900 MHz, delivering 51.1 W output power with 78 per cent peak drain-efficiency (Gustavsson *et al.*, 2007).

A highly efficient class-F power amplifier (PA) using a GaN HEMT designed at W-CDMA band of 2.14 GHz has the drain efficiency and power-added efficiency of 75.4 per cent and 70.9 per cent with a gain of 12.2 dB at an output power of 40.2 dBm (Lee *et al.*, 2008). The GaN HEMTs have also proven to be very attractive and viable as a power source for millimeter wave applications (Kanamura *et al.*, 2004; Streit *et al.*, 2005; Micovic *et al.*, 2004; Darwish *et al.*, [IEEE Trans. Microwave Theory and Techniques]; Wu *et al.*, 2005). Similar to microwave frequencies, microstrip and CPW

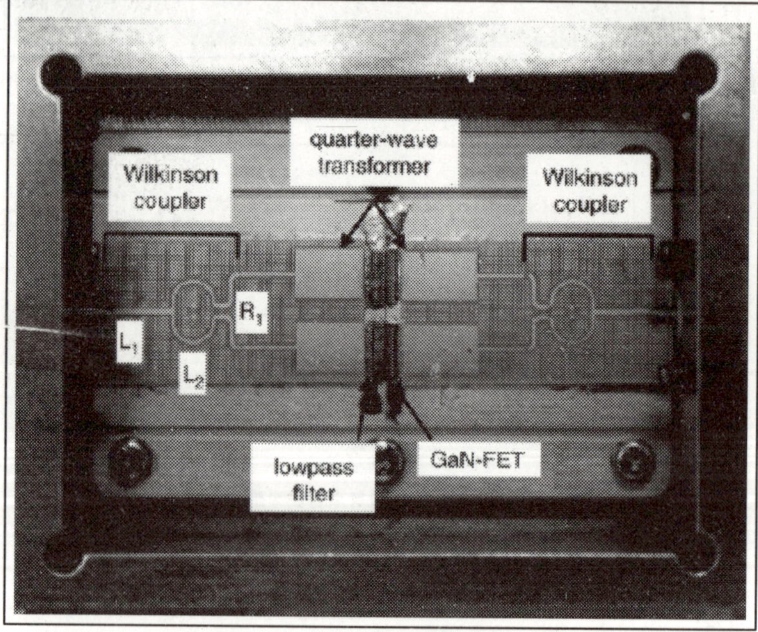

Figure 1.10: Top View of Developed GaN-FET Amplifier (Okamoto *et al.*, 2007)

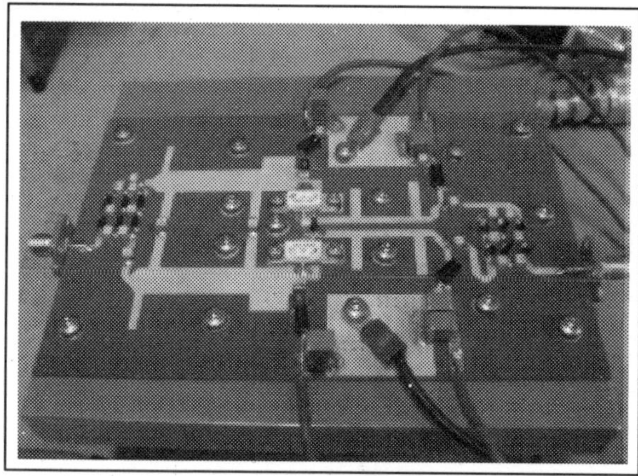

Figure 1.11: A Photo of GaN MESFET Power Amplifier (Gustavsson *et al.*, 2007)

MMICs have been demonstrated, a microstrip Ka band GaN MMIC power amplifier capable of delivering 11 W of output power (Kanamura *et al.*, 2004). Wu *et al.*, announced an amplifier with a 1.5-mm-wide device produced 8.05 W output power at 30 GHz with 31 per cent PAE and 4.1 dB associated gain.

The output power matches that of a GaAs-based MMIC with a 14.7-mm-wide output device but with a 10 times smaller size. Recently, GaN MMIC performance has been demonstrated at W-band as well (Wu *et al.*, 2005). The demonstrated high power amplifiers and MMICs with high power densities, efficiencies and suitable gain will enable the proliferation of solid state solutions at millimeter wave frequencies.

SiC Based Power Amplifiers

Silicon carbide (SiC) Metal-Semiconductor Field Effects Transistors (MESFET) has the operational frequency limitation of X-band. This is the reason that they have been mostly employed in microwave, broadband radio frequency (RF), radar and wireless communication power amplifiers.

The efficiency improvement of power amplifiers decreases power consumption and heat sink requirements, and increases output power because power amplifiers account for the majority of power consumption in wireless communications. Therefore, the switching-mode power amplifiers have recently received attention to improve efficiency of power amplifiers. A high efficiency, class-E RF power amplifier in the VHF range is implemented. A maximum drain DC to RF efficiency of 87 per cent was predicted and 86.8 per cent achieved with 20.5 W output power at a 30 V drain voltage (Figures 1.13 and 1.14). The SiC MESFETs used appear to offer significant advantages, particularly for space applications, over gallium arsenide (GaAs) transistors, which are inherently low voltage devices and more difficult to operate in class-E due to the high drain peak voltage occurring in this class of operation(Franco and Katz, 2007).

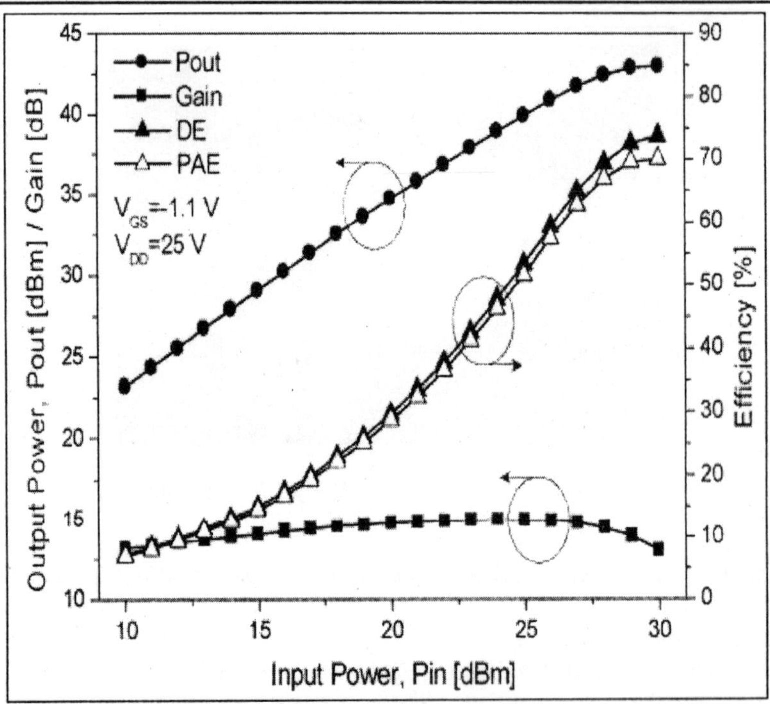

Figure 1.12a: Measured Output Power, Gain, and Efficiency of the Class-E Power Amplifier Using GaN HEMT (Ui *et al.*, 2007)

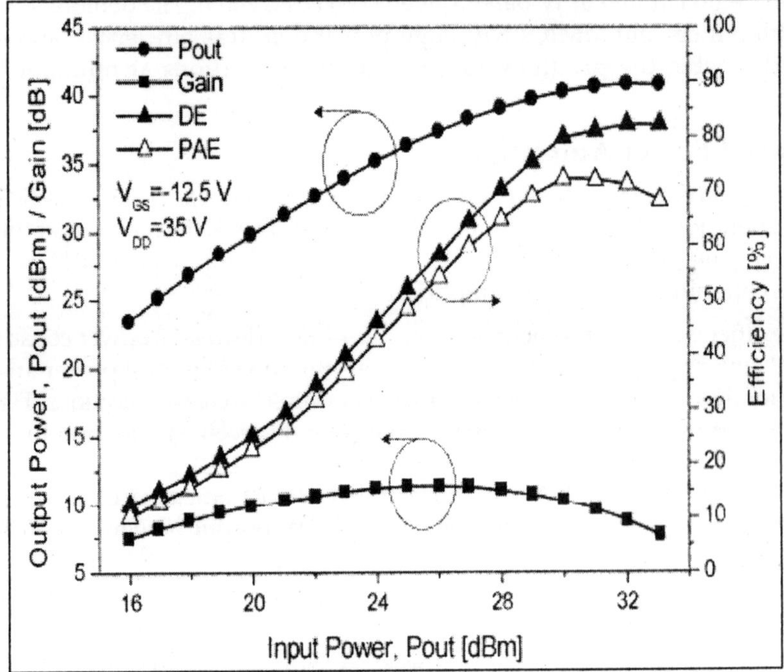

Figure 1.12b: Measured Output Power, Gain, and Efficiency Characteristics of the Class-E Power Amplifier Using SiC MESFET, at a Single Tone of 2.14 GHz (Ui *et al.*, 2007)

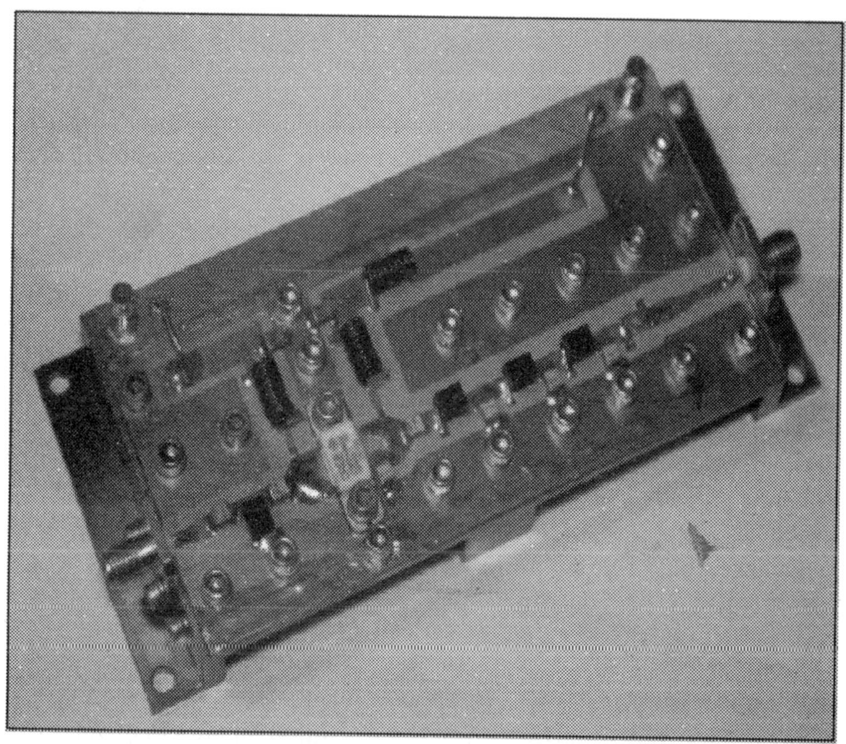

Figure 1.13: High Efficiency SiC MESFET Class-E VHF Amplifier (Franco and Katz, 2007)

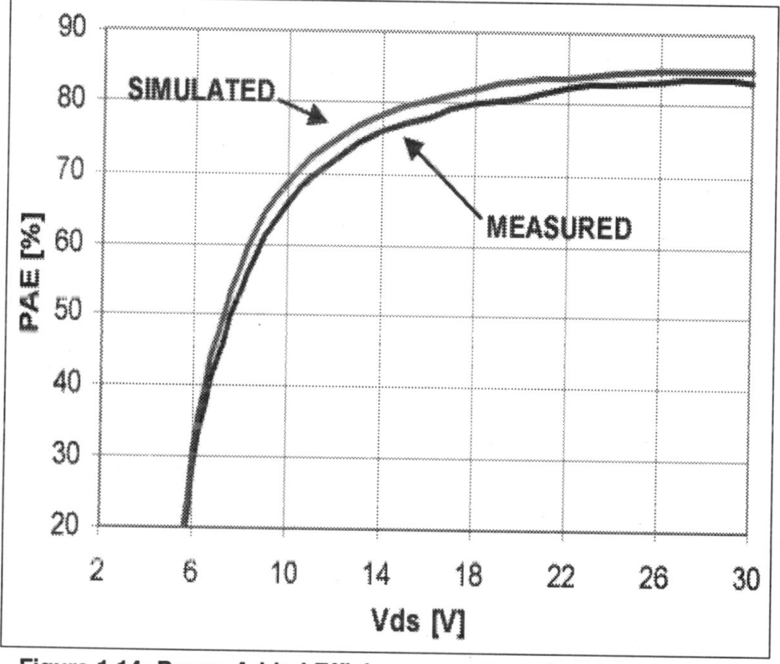

Figure 1.14: Power Added Efficiency as a Function of Drain Voltage (Franco and Katz, 2007)

Another class-E power amplifier using SiC MESFET is reported with power added efficiency (PAE) of 72.3 per cent and a gain of 10.3 dB at an output power of 40.3 dBm, through significant reduction of harmonic power levels (Figure 1.12) (Lee and Jeong, 2007). A new type of pulse input class C power amplifier is reported with a maximum PAE of 80 per cent at 500 MHz, a gain of 36.9 dB and power density of 1.07 W/mm (Azam *et al.*, 2007).

A power amplifier (PA) for WiMAX military applications in Nato Band I (225 to 400 MHz) has been simulated, assembled and tested. Under 802.16 OFDM 64QAM3/4 modulation, the average output power is 25W throughout the bandwidth (Figures 1.15 and 1.16) (Risso *et al.*, 2007). A class-E power amplifier using a SiC MESFET is designed and tested at 2.14 GHz (Figure 1.17). The peak power-added efficiency of 72.3 per cent with a power gain of 10.27 dB is achieved at an output power of 40.27 dBm (Lee and Jeong, 2007).

Most publications on SiC microwave components concerns L to C-band operation, since these are important radar frequency bands. However, the frequency performance up to the X band has been predicted to be good (Sudow *et al.*, 2006) and SiC MESFETs with power densities of 4.5 W/mm at 10 GHz have been demonstrated (Trew, 1997). The power amplifier in a narrow-band 3–3.5 GHz, designed based on a 6-mm SiC MESFET. The amplifier was measured on-wafer and showed a typical power gain of 7 dB, an output power of 2.5 W in continuous wave (CW) operation, and 8W in pulsed mode operation at 3 GHz (Figures 1.18 and 1.19) (Sadler *et al.*, 2000).

A single-stage 26 W negative feedback power amplifier is implemented, covering the frequency range 200-500 MHz using a 6 mm gate width SiC Lateral Epitaxy

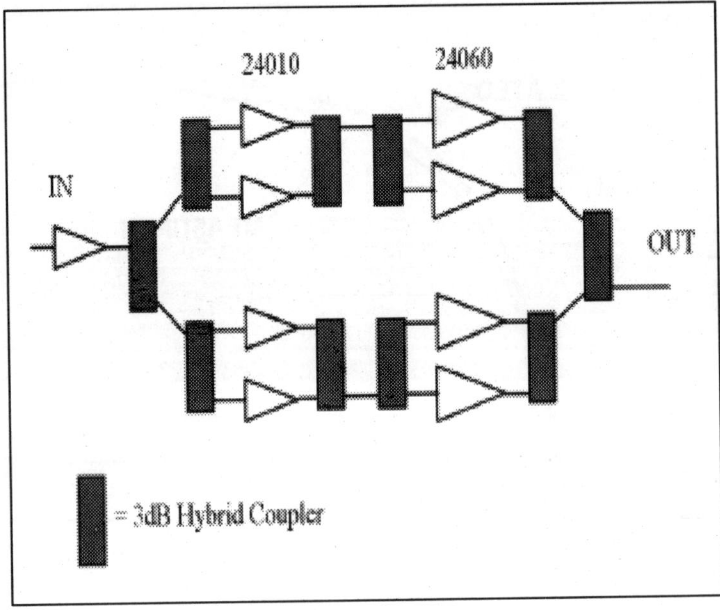

Figure 1.15: Amplifier Topology (Risso *et al.*, 2007)

Figure 1.16: PA Orototype (Risso *et al.,* 2007)

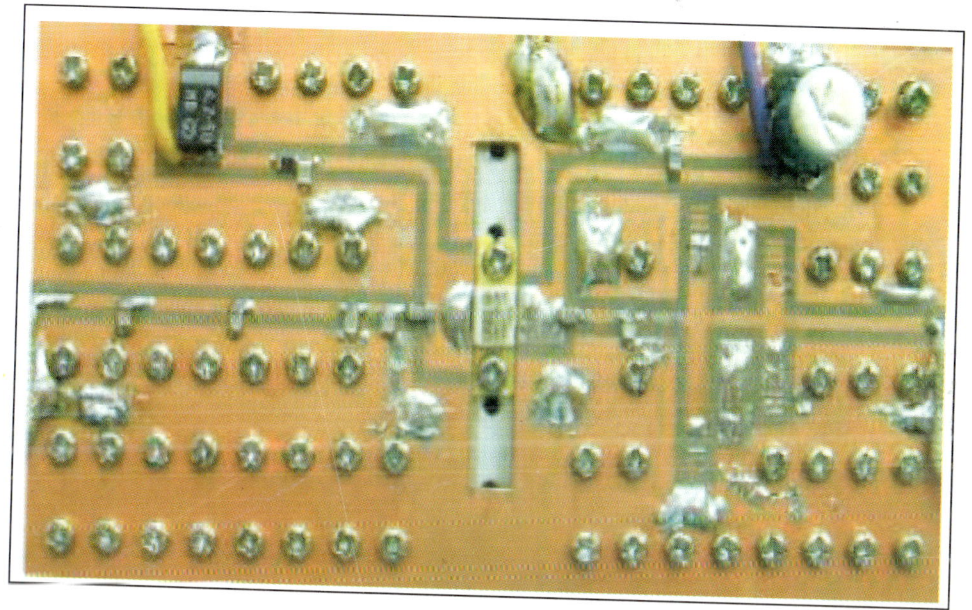

Figure 1.17: Photograph of a Fabricated Class-E SiC MESFET Power Amplifier (Lee and Jeong, 2007)

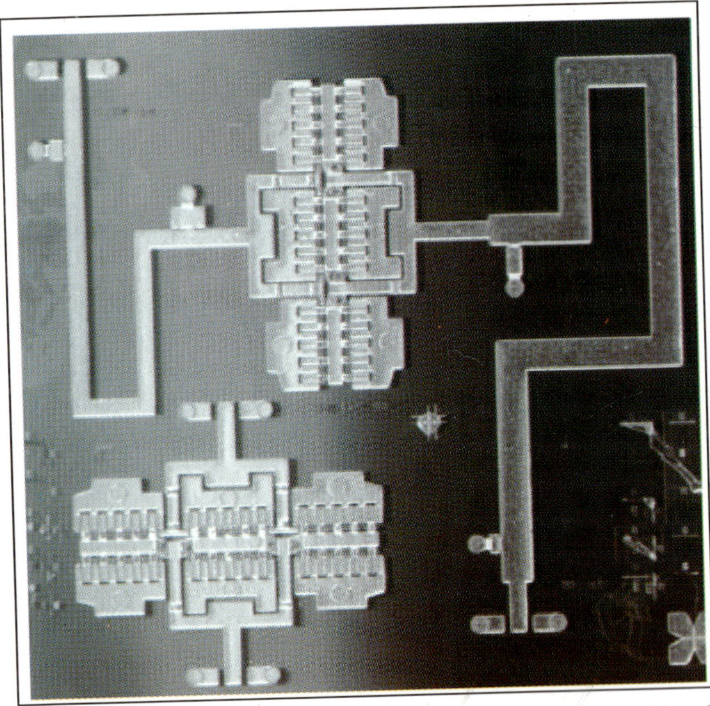

Figure 1.18: 3-GHz SiC MESFET-based Power Amplifier Together with a Cut-out of the MESFET Transistor (seen in the lower left corner) (Sadler *et al.*, 2000)

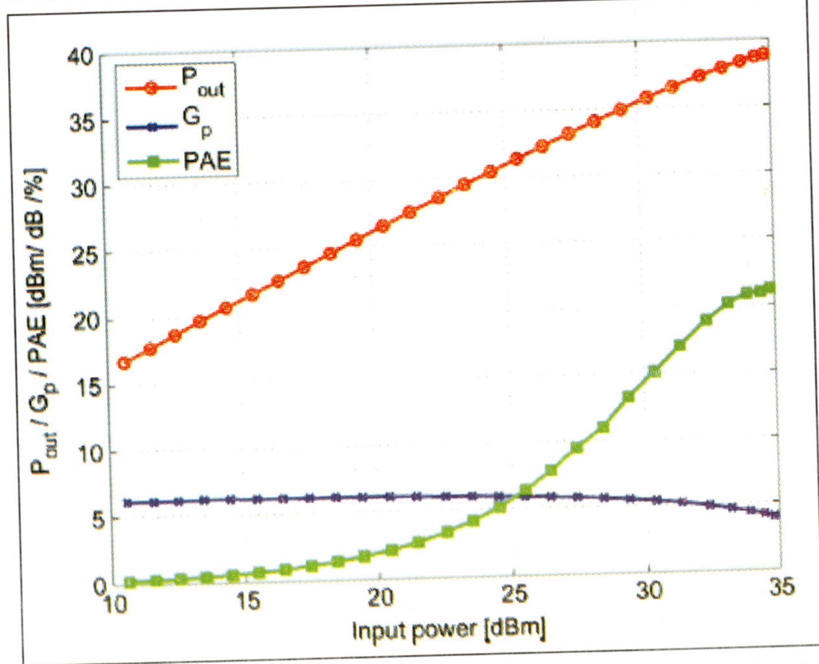

Figure 1.19: Pulsed Power Sweep at 3 GHz of the MMIC PA Showing an Unsaturated Output Power of 8 W (Sadler *et al.*, 2000)

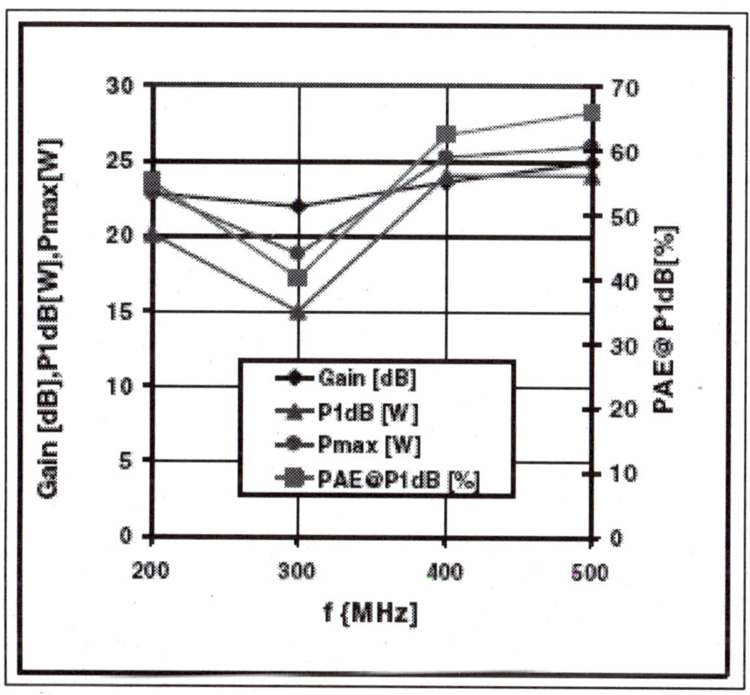

Figure 1.20: Measured Result of Gain, P 1dB and PAE at P 1dB and Pmax Versus Frequency at 60 V (Azam *et al.*, 2006)

MESFET. The results at 60 V drain bias at 500 MHz are, 24.9 dB power gain, 44.15 dBm output power (26 W) and 66 per cent PAE (Figure 1.20) (Azam *et al.*, 2006). Previous reports on SiC MESFET transistor amplifiers designed for frequencies below 500 MHz showed an output power of 37.5 W and a power density of 1.78 W/mm, the gain was 8 dB, and the efficiency in class A-AB was 55 per cent at 500 MHz. The IMD3 level at 10 dB back-off from P_{1dB} was–35 dBc for a 1 MHz frequency offset between tones (Villard *et al.*, 2003).

A Class C mode power amplifier is implemented with a 75 V power supply voltage. The total output power was measured to be 2100 W with a power gain of 6.3 dB, collector efficiency of 45 per cent and power added efficiency of 35 per cent . This is the first time; SiC BJTs have been used to produce an output power in excess of 2 kW at 425 MHz. Although the gain and PAE are not very high, the individual cells are capable of producing 50 W with a gain of 9.3 dB and 51 per cent collector efficiency (Agarwal *et al.*, 2005). Two SiC MESFET package and prototype power amplifier were demonstrated with P_{1dB} output power of 26W and 35W, respectively. High output power and gain were maintained through L-band operation across 500 MHz bandwidth (950MHz to 1500 MHz with 10 dB gain) for the SiC PA module, which will be a critical challenge for other semiconductor devices (Chen *et al.*, 2004).

These results show that although SiC based PAs presently can not compete with GaN and other conventional devices like GaAs in terms of frequency but in terms of power and efficiency, they could be the strong competitors and future devices for, RADAR, Electronic Warfare (EW), Wireless Communications and base stations applications.

Conclusion

In several RF applications, as well as in radar and military systems, the development of circuits and sub-systems with broadband capabilities is required, which requires wide band gap materials (like, GaN and SiC) due to their superior properties of high power, high frequency, high breakdown voltage and high temperature operation. GaN and SiC-based RF power devices and power amplifiers have made substantial progresses. They have the potential as future devices for, RADAR, Electronic Warfare (EW), Wireless Communications, base stations and other electronics applications.

References

Azam, Sher, Jonsson, R., Wahab, Q., 2006. "Single-stage, High Efficiency, 26-Watt power Amplifier using SiC LE-MESFET" IEEE Asia Pacific Microwave Conf. (APMC-2006), pp.441–444.

Agarwal, Anant, Haley, Jeremy, Bartlow, Howard, McCalpin, Bill, Capel, Craig, Palmour, John W., 2005. "2100 W at 425 MHz with SiC RF Power BJTs" 63rd Device Research Conference, pp. 189–190.

Azam, Sher, Svensson, C., Wahab, Q., 2007. "Designing of High Efficiency Power Amplifier Based on Physical Model of SiC MESFET in TCAD." IEEE Proceedings of International Conference on Applied Sciences and Technology Islamabad, Pakistan, pp. 40-43.

Chen, P., Chang, H.R., Li, X., Luo, Ben, 2004. "Design and fabrication of SiC MESFET Transistor and broadband power amplifier for RF applications", Proceedings of 2004 International Symposium on Power Semiconductor Devices and ICs, Kitakyushu, pp.317–318.

Colantonio, P., Giannini, F., Giofre', R., Piazzon, L., 2008. "High-efficiency ultra-wideband power amplifier in GaN technology", Electronic Letters Vol. 44 No. 2.

Darwish, A., Boutros, K., Luo, B., Huebschman, B. D., Viveiros, E., Hung, H. A. "AlGaN/GaN Ka-band 5-W MMIC amplifier", IEEE Trans. Microwave Theory and Techniques, to be published.

Franco, Marc, Katz, Allen, 2007. "Class-E Silicon Carbide VHF Power Amplifier", Microwave Symposium. IEEE/MTT-S International, pp. 19-22.

Gassmann, J., Watson, P., Kehias, L., Henry, G., 2007. "Wideband, high efficiency GaN power amplifiers utilizing a non-uniform distributed topology", IEEE MTT-S Sym. Digest, pp. 615–618.

Gassmann, J., Watson, P., Kehias, L., Henry, G., 2007. "Wideband, High-Efficiency GaN Power Amplifiers Utilizing a Non-Uniform Distributed Topology ", Microwave Symposium, IEEE MTT-S International, pp. 615-618.

Gustavsson, Ulf, Lejon, Thomas, Fager, Christian, Zirath, Herbert, October 2007. "Design of highly efficient, high output power, L-band class D-1 RF power amplifiers using GaN MESFET devices", Proceedings of the 2nd European Microwave Integrated Circuits Conference, Munich Germany, pp. 291-294.

Hong, Sungchul, Woo, Young Yun, Kim, Ildu, Kim, Jangheon, Moon, Junghwan, Kim, Han Seok, Lee, Jong Sung, Kim, Bumman, 2007. "High Efficiency GaN HEMT Power Amplifier optimized for OFDM EER Transmitter", Microwave Symposium, IEEE MTT-S International, pp. 1247-1250.

Jyomasa', Kazuhiro, Yamanaka', Koji, Mori, Kazutomi, Noto, Hifumi, Ohtsuka, Hiroshi, Nakayama, Masatoshi, Yoneda, Satoshi, Kamo, Yoshitaka, Isota, Yoji, 2007. "GaN HEMT 60W Output Power Amplifier with Over 50 per cent Efficiency at C-Band 15 per cent Relative Bandwidth Using Combined Short and Open Circuited Stubs", IEEE MTT-S Digest, pp.1255-1258.

Kikkawa, T., Maniwa, T., Hayashi, H., Kanamura, M., Yokokawa, S., Nishi, M., Adachi, N., Yokoyama, M., Tateno, Y., Joshin, K., 2005. "An over 200-W output power GaN HEMT push-pull amplifier with high reliability" IEEE MTT-S Digest, Vol. 3, pp. 1347–1350.

Kikkawa, T., 2004. "Recent progress and future prospects of GaN HEMTs for base-station applications", IEEE Compound Semiconductor Integrated Circuit Symposium, pp. 17–20.

Kawano, A., Adachi, N., Tateno, Y., Mizuno, S., Ui, N., Nikaido, J., Sano, S., 2005. "High-efficiency and wide-band single-ended 200 W GaN HEMT power amplifier for 2.1 GHz W-CDMA base station application", Asia-Pacific Microwave Conference (APMC) Proceedings, Vol. 3, pp. 4–7.

Kim, Jangheon, Moon, Junghwan, Woo, Young Yun, Hong, Sungchul, Kim, Ildu, Kim, Jungjoon, Kim, Bumman, 2008. "Analysis of a Fully Matched Saturated Doherty Amplifier With Excellent Efficiency", IEEE Transaction on MTT, Vol. 56, NO. 2.

Kanamura, M., Kikkawa, T., Joshin, K., 2004. "A 100-W high-gain AlGaN/GaN HEMT power amplifier on a conductive n-SiC substrate for wireless base station applications", IEEE IEDM Technical Digest, pp. 799–802.

Lee, Yong-Sub, Lee, Mun-Woo, Jeong, Yoon-Ha, 2008. "High-Efficiency Class-F GaN HEMT Amplifier With Simple Parasitic Compensation Circuit", IEEE Microwave and Wireless Components Letters, Vol. 18, No. 1.

Lee, Yong-Sub, Jeong, Yoon-Ha, 2007. "Applications of GaN HEMTs and SiC MESFETs in High Efficiency Class-E Power Amplifier Design for WCDA Applications", [Microwave Symposium, IEEE, MTT-S International, pp. 615-618.

Lee, Yong-Sub, Jeong, Yoon-Ha, 2007. "A high efficiency Class-E power amplifier using SiC MESFET" Microwave and optical technology Letters, Vol. 49, No. 6.

Mishra, Umesh K., Fellow IEEE, Shen, Likun, Kazior, Thomas E. and Wu, Yi-Feng, February 2008. "GaN-Based RF Power Devices and Amplifiers", Proceedings of the IEEE Vol. 96, No. 2.

Maekawa, A., Yamamoto, T., Mitani E., Sano, S., 2006. "A 500W Push Pull AlGaN/GaN HEMT Amplifier for L-Band High Power Application", IEEE MTT-S Digest pp. 722-725.

Maekawa, A., Nagahara, M., Yamamoto, T., Sano, S., 2005. "100 W high-efficiency GaN HEMT amplifier for S-band wireless system", In Gallium Arsenide and Other Semiconductor Application Symposium, pp. 497–500.

Moon, Junghwan, Kim, Jangheon, Kim, Ildu, Kim, Jungjoon and Kim, Bumman, 2008. "A Wideband Envelope Tracking Doherty Amplifier for WiMAX Systems", IEEE Microwave and Wireless component Letters, Vol. 18, No. 1.

Micovic, M., Kurdoghlian, A., Hashimoto, P., Hu, M., Antcliffe, M., Willadsen, P. J., Wong, W. S., Bowen, R., Milosavljevic, I., Schmitz, A., Wetzel, M., Chow, D. H., 2006.BGaN HFET for W-band power applications", IEEE International Electron Devices Meeting.

Micovic, M., Kurdoghlian, A., Moyer, H. P., Hashimoto, P., Schmitz, A., Milosavjevic, I., Willadesn, P. J., Wong, W.-S., Duvall, J., Hu, M., Delaney, M. J. and Chow, D. H., 2004. "Ka-band MMIC power amplifier in GaN HFET technology", IEEE MTT-S International Microwave Symposium Digest, Vol. 3, pp. 1653–1656.

Okamoto, Y., Nakayama, T., Ando, Y., Wakejima, A., Matsunaga, K., Ota, K., Miyamoto, H., 2007. "230 W C-band GaN-FET power amplifier", Electronics Letters Vol. 43, No. 17, pp. 1-2.

Otsuka, H., Mori, K., Yukawa, H., Minamide, H., Kittaka, Y., Tsunoda, T., Ogura, S., Ikeda, Y., Takagi, T., June 2004. "Over 65 per cent efficiency 300 MHz bandwidth C-band internally-matched GaAs FET designed with a large-signal FET model," IEEE MTT-S Digest, pp.521-524.

Risso, Luca, Armoni, Alberto, Petacchi, Luca, 2007. "A 225-400MHz WiMAX 20W SiC Power Amplifier", Proceedings of the 37th European Microwave Conference, pp. 1291-1294.

Streit, D. C., Gutierrez-Aitken, A., Wojtowicz, M., Lai, R., 2005. "The future of compound semiconductors for aerospace and defense applications", Compound Semiconductor Integrated Circuit Symposium, (CSIC-05), pp. 4.

Sheppard, S. T., Smith, R. P., Pribble, W. L., Ring, Z., Smith, T., Allen, S. T., Milligan, J., Palmour, J.W., 2002. "High power hybrid and MMIC amplifiers using wide-bandgap semiconductor devices on semi-insulating SiC substrates", 60th Device Research Conference, (DRC-2002). Conference Digest, Jun. 24–26, pp. 175–178.

Sudow, Mattias *et al.*, 2006. "An SiC MESFET-Based MMIC Process," IEEE Transactions on Microwave Theory and Techniques, Vol. 54, No. 12, pp. 4072–4079.

Sadler, R., Allen, S., Pribble, W., Alcorn, T., Sumakeris, J., Palmour, J., 2000. "SiC MESFET hybrid amplifier with 30-W output power at 10 GHz," Proc. IEEE Cornell Conf. High Performance Devices, pp. 173–177.

Tayrani, R., 2007. "A Spectrally pure 5.0 W, High PAE, (6-12 GHz) GaN Monolithic Class E Power Amplifier for Advanced T/R Modules", IEEE Radio Frequency Integrated Circuits Symposium, pp. 581-584.

Trew, R. J., 1997. "Experimental and simulated results of SiC microwave power MESFETs," Phys. Stat. Sol. A, Vol. 162, No. 1, pp. 409–419.

Ui., Norihiko, Sano, Hiroaki, Sano, Seigo, 2007. "A 80W 2-stage GaN HEMT Doherty Amplifier with–5OdBc ACLR, 42 per cent Efficiency 32dB Gain with DPD for W-CDMA Base station", IEEE MTT-S Digest, pp. 1259-1262.

Villard, F., Prigent, J.-P., Morvan, E., Dua, C., Brylinski, C., Temcamini, T., Pouvil, P., 2003. "Trap-Free Process and Thermal Limitations on Large- Periphery SiC MESFET for RF and Microwave Power", IEEE Trans MTT, Vol. 51, pp.1129 – 1134.

Wu, Y.-F., Moore, M., Abrahamsen, A., Jacob-Mitos, M., Parikh, P., Heikman, S. and Burk, A., 2007. "High-voltage Millimeter-Wave GaN HEMTs with 13.7 W/mm Power Density", IEEE Electron Devices Meeting, IEDM 2007. IEEE International, pp. 405-407.

Wu, Y.-F., Moore, M., Saxler, A., Wisleder, T., Mishra, U. K., Parikh, P., 2005. "8-watt GaN HEMTs at millimeter wave frequencies, IEEE International Electron Devices Meeting, (IEDM) Technical Digest, pp. 583–585.

Wu, Y.-F., Saxler, A., Moore, M., Smith, R. P., Sheppard, S., Chavarkar, P. M., Wisleder, T., Mishra, U. K., Parikh, P., 2004. "30-W/mm GaN HEMTs by Field Plate Optimization", IEEE Electron device Lett., Vol. 25, No. 3, pp 117-119.

Wakejima, A., Nakayama, T., Ota, K., Okamoto, Y., Ando, Y., Kuroda, N., Tanomura, M., Matsunaga, K., Miyamoto, H., 2006. "Pulsed 0.75kW output single-ended GaN-FET amplifier for L/S band applications", Electronics Letters, Vol. 42, No. 23.

Wataru, Saito, Tomokazu, Domon, Ichiro, Omura, Masa-hiko, Kuraguchi, Yoshiharu, Takada, Kunio, Tsuda *et al.*, 2006. "Demonstration of 13.56-MHz Class-E amplifier using a high-voltage GaN power-HEMT", IEEE Electron Dev Lett Vol. 27 No. 5, pp 326–8. 235.

Wu, Y.-F., Wood, S. M., Smith, R. P., Sheppard, S., Allen, S. T., Parikh, P., Milligan, J., 2006. "An internally-matched GaN HEMT amplifier with 550-watt peak power at 3.5 GHz", IEEE International Electron Devices Meeting (IEDM).

Yamanaka, K., Mori, K., Iyomasa, K., Ohtsuka, H., Noto, H., Nakayama, M., Kamo, Y., Isota, Y., 2007. "C-band GaN HEMT Power Amplifier with 220W Output Power", Microwave Symposium, IEEE/MTT-S International, pp. 1251-1254.

Yamanaka, K., Iyomasa, K., Ohtsuka, H., Nakayama, M., Tsuyama, Y., Kunii, T., Kamo, Y., Takagi, T., 2005. "S and C band over 100W GaN HEMT 1-chip high power amplifiers with cell division configuration", European Microwave Conf., in Gallium Arsenide and Other Semiconductor Application Symposium, (EGAAS 2005), pp. 241–244.

Chapter 2

MEMS for Investigating Mechanical Behavior of Nanoscale Metal Films

Jagannathan Rajagopalan and M. Taher A. Saif*

Department of Mechanical Science and Engineering,
University of Illinois at Urbana-Champaign, Urbana, IL 61820, USA
**E-mail: saif@uiuc.edu*

ABSTRACT

Thin metal films exhibit mechanical properties that are vastly different from their bulk, coarse grained counterparts. However, mechanical testing of nanoscale thin films has been a challenge due to various reasons. For example, because of the small size of these samples it becomes imperative to measure the forces and displacements with a high degree of precision. Micro-electro-mechanical systems (MEMS), which have the capability to measure small forces and displacements, are hence ideal platforms for testing thin films. In this paper, we provide a brief overview of thin film testing and discuss the unexpected recovery of plastic deformation in nanocrystalline aluminum and gold thin films.

Keywords: *Micro-electro-mechanical systems, Thin metal films, Nanoscale thin films, Nanocrystalline aluminum, Gold thin films.*

Introduction

Thin metal films have found increasing application in integrated circuits and MEMS based devices and this has led to much interest in investigating their mechanical properties, as the structural integrity of these films is vital to the reliability of these circuits and devices. It is also well known that the properties of thin metal films differ substantially from the bulk properties (Artz, 1998). However, mechanical testing of these thin films has been a challenge because of the small size (Madou, 1997) as well as the fragile nature of the specimens. Some of the testing methods developed to study thin films include wafer curvature technique, bulge testing, nano-indentation, and uniaxial tensile testing. In wafer curvature measurement technique (Flinn *et al.*, 1987) the thermal mismatch between the film and the substrate is used to induce strain in the thin film by raising the temperature. The stress in the film is then calculated from the curvature change. Nano-indentation (Nix, 1989) measures mechanical properties by indenting thin film samples on a substrate. However, in both these techniques interpreting the stress-strain data is complicated because of the presence of the substrate.

Bulge testing, which uses a pressure difference across a thin membrane to strain the membrane, is one way to measure the properties of freestanding films (Xiang *et al.*, 2005) .Stress and strain are calculated from analytical expressions that relate them to the pressure difference across the membrane and its deflection. Uniaxial tensile testing, on the other hand provides a direct measure of stress and strain in the sample. However, care needs to be taken to ensure proper gripping of the sample and alignment of the specimen with the loading direction. Several other techniques also exist for thin film testing. In many of these techniques a load cell is used to measure the force on the specimen and the sample is deformed by some type of micro actuation. The strain is measured in various ways including using optical interference from photoresist islands on the sample (Ruud *et al.*, 1993) and metal markers (Yuan and Sharpe, 1997).

The aforementioned techniques allow one to obtain the macroscopic response of thin films but they do not allow simultaneous microstructural characterization. This was first made possible by a MEMS based tensile testing device developed by Haque and Saif (Haque and Saif, 2004) .This device, because of its small dimensions, could be placed in analytical chambers such as TEM and one could simultaneously observe the microstructural evolution as well as the macroscopic stress and strain. Furthermore, as the thin film specimen and the testing device are co-fabricated it circumvents the problems associated with specimen gripping, and ensures perfect alignment of the specimen with the loading axis. A modified version of this device, incorporating a new fabrication procedure and design improvements, was developed by Han and Saif (Han and Saif, 2006) . Figure 2.1 shows one of these tensile testing devices and describes its operation.

This tensile testing device was used to conduct uniaxial loading and unloading experiments on nano-crystalline aluminum and gold thin films which showed that, contrary to expectation, the plastic deformation in these films was largely (50-100 per cent) recoverable (Figure 2.2). Furthermore, this recovery was time dependent and thermally activated (Rajagopalan *et al.*, 2007). Since this recovery occurred at a macroscopically stress free state, the driving force for the recovery has to come from internal stresses. One likely cause of such internal stresses is inhomogeneous deformation of the specimen.

In any polycrystalline metal there is normally a distribution of grain sizes and the individual grains yield at slightly different stress levels. This distribution of grain sizes is seen in nano-crystalline materials as well. However, if the grain size becomes very small it becomes energetically unfavorable to sustain dislocation plasticity and

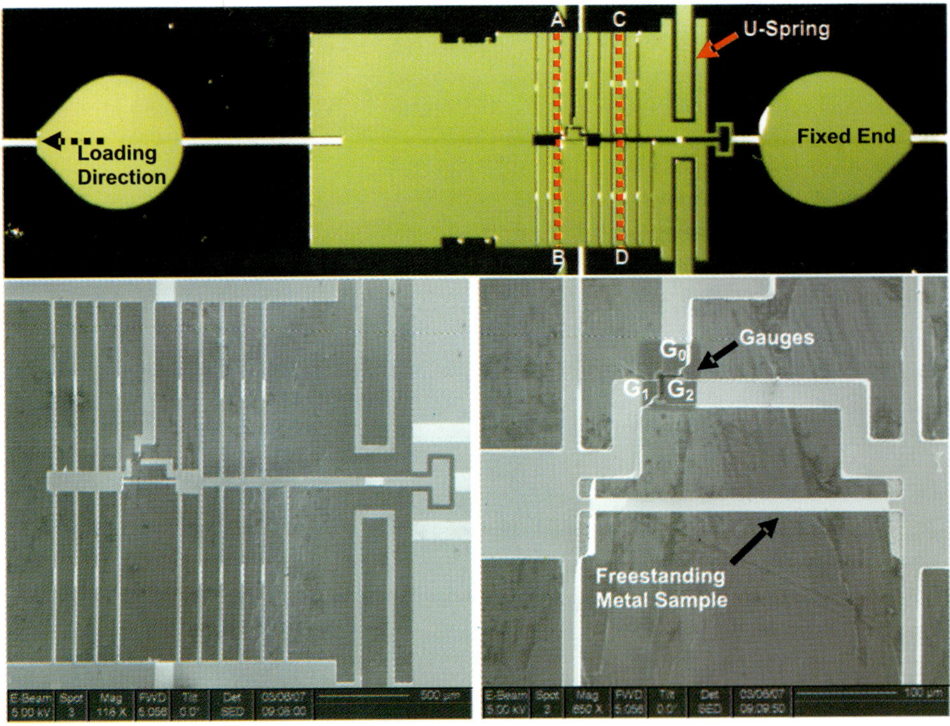

Figure 2.1

Top: Optical micrograph of a typical stage used for tensile testing. AB–force sensing beams, CD–support beams. When the stage is loaded, the U-springs deform and apply uniaxial tension on the sample. Support beams and U-springs ensure alignment of the sample with the loading axis. Bottom left: Scanning electron micrograph of the sample area in a typical stage. Bottom right: A higher magnification shot of the sample.

During loading, the deflection of the force sensing beams, provided by the relative displacement of gauge G_1 with respect to fixed gauge G_0, multiplied by their stiffness gives the force on the sample. The elongation of the sample is given by the relative displacement of G_2 with respect to G_1.

the deformation in these grains remains elastic till very high stress levels. Hence in nano-crystalline metals, the grains at the lower end of the distribution are primarily likely to deform elastically where as larger grains deform plastically. This would lead to lower stresses in the larger grains and high stresses in the smaller grains, that is, an inhomogeneous stress distribution.

This inhomogeneous stress distribution leads to residual stresses upon unloading. To reduce these residual stresses the larger grains undergo reverse plastic deformation which leads to recovery of plastic deformation. A deformation model based on a thermally activated dislocation propagation mechanism which considers such inhomogeneous deformation has been shown to reasonably explain this strain recovery ('On Plastic strain recovery in freestanding nanocrystalline metal films by

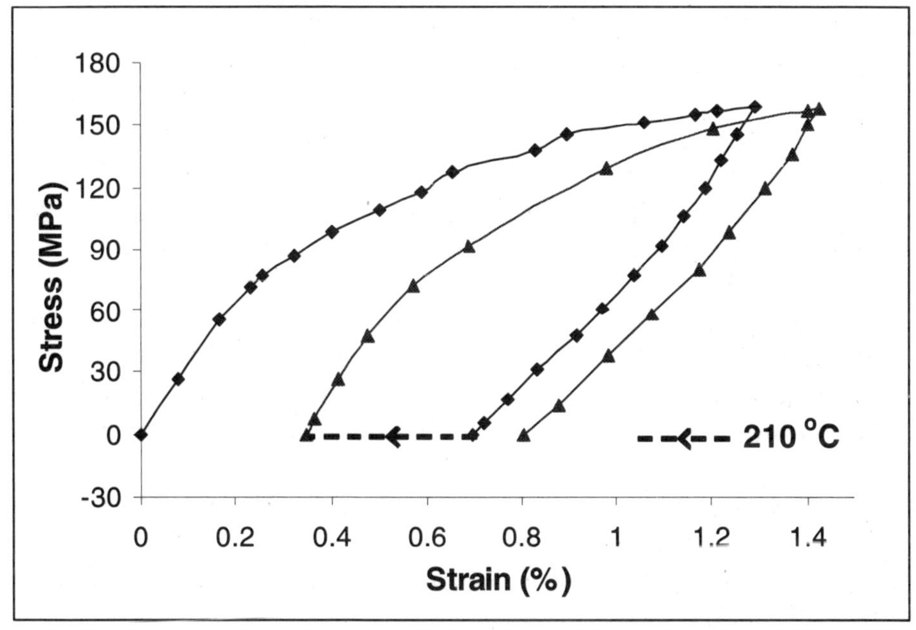

Figure 2.2: Stress-strain Curve of 200 nm Thick Gold Film with a Grain Size of 50 nm (Rajagopalan *et al.*, 2007). The dashed arrow represents the strain recovered upon annealing at 210ºC.

Rajagopalan *et al.*). A different model that takes into account the inhomogeneous nature of deformation, but based on a different microscopic mechanism, has also been shown to result in strain recovery (Wei *et al.*, 2007).

In conclusion, MEMS based devices serve as excellent platforms to study the mechanical behavior of nano-scale materials. They provide highly precise force and displacement measurement capabilities and because of their small size allow testing in analytical chambers such as TEM.

References

Artz, E., 1998. "Size effects in materials due to microstructural and dimensional constraints: a comparative review," Acta Mater., 46(16), pp.5611-5626.

Flinn, P. A., Gardner, D. S., Nix, W. D., 1987. "Measurement and interpretation of stress in aluminum-based metallization as a function of thermal history," IEEE Trans. Electron Devices, 3, pp.689-699.

Haque, M. A., Saif, M. T. A., 2004. "Deformation mechanisms in free-standing nanoscale thin films: a quantitative in situ transmission electron microscope study," Proc. Natl. Acad. Sci., 101(17), pp.6335-6340.

Han, J. H., Saif, M. T. A., 2006. "In situ microtensile stage for electromechanical characterization of nanoscale freestanding films," Rev. Sci. Inst., 77(4), pp. 45102-1-8.

Madou, M. J., 1997. "Fundamentals of microfabrication," CRC Press.

Nix, W. D, 1989. "Mechanical properties of thin films," Metall. Trans., 20A, pp. 2217-2245.

Ruud, J. A., Josell, D., Spaepen, F., 1993. "A new method for tensile testing of thin films," J. Mater. Res., 8(1), pp.112-117.

Rajagopalan, J., Han, J. H., Saif, M. T. A., 2007. "Plastic deformation recovery in freestanding nanocrystalline aluminum and gold thin films," Science, 315, pp.1831-1834.

Rajagopalan, J., Han, J. H., Saif, M. T. A. "On Plastic strain recovery in freestanding nanocrystalline metal films," Scr. Mater, (accepted for publication).

Wei, Y., Bower, A. F., Gao, H., 2007. "Recoverable creep deformation due to heterogeneous grain-boundary diffusion and sliding," Scr. Mater, 57, pp.933-936.

Xiang, Y., Chen, X. and Vlassak, J. J., 2005. "The plane-strain bulge test for thin films," J. Mater. Res., 20(9), pp.2360-2370.

Yuan, B., Sharpe, W. N., 1997. "Mechanical testing of polysilicon thin films with the ISDG," Expt. Tech., 21(2), pp.32-35.

Chapter 3

A Vision on Future of Consumer Electronics: Some System Architecture and Technology Consequences

Aly Aamer Syed
NXP Semiconductors, Corporate I&T/Research,
High Tech Campus 32, The Netherlands
E-mail: aly.syed@nxp.com

ABSTRACT

In ambient intelligence vision of the future, electronic devices are embedded in the environment of the user such as the furniture and the building and cater to users needs in an intuitive manner. Today, the world of consumer electronics is moving from stand-alone devices such as a TV to a networked system formed by a number of devices that are available to the user. Consider for example the TV, the video recorder, the PC and the mobile phone etc. These devices, which are present in many homes even today, can form a network and complement each other's functions. This move towards networked systems is often seen as a step towards the vision of Ambient Intelligence. This paper introduces the vision of ambient intelligence and presents some technological challenges that need to be overcome to make the vision of Ambient Intelligence come true.

Keywords: Consumer electronics systems, Ambient intelligence, System architecture, Pervasive computing technology, Nanometer science.

Introduction

From earliest times humans have always had a need to be entertained, communicate, be informed and be creative. Technology enables these needs to be met in different ways opening new possibilities that were not thought possible before. In 1900 people would not have believed that technology would one day make it possible for everyone to have a cell phone allowing them to communicate whenever and wherever they wanted. For the future, the visionaries sketch visions, which people call Ambient Intelligence, Pervasive Computing, Ubiquitous computing, Sentient computing etc. All these visions pose some requirements on technologies that will be needed and also some issues that need to be resolved to make these visions come true. In this paper, we discuss briefly these visions and present a non-exhaustive list of issues, as seen from system architecture viewpoint, which need resolution. The goal is to draw attention of scientists whose findings will eventually form the new technologies to be aware of these issues.

Views on the Future

Philips (www.philips.com) have a vision that, in the year 2020, people will relate to electronics in more natural and comfortable ways in comparison to the present situation. They believe that current inventions, by Philips and others, will make electronics 'smart'. Technological breakthroughs will also allow us to integrate 'smart electronics' into more friendly environments. In their vision of 'Ambient Intelligence': people live with ease in digital environments in which the electronics are sensitive to people's needs, personalized to their requirements, anticipatory of their behavior and responsive to their presence.

Another vision that goes in the same direction is that of Pervasive computing. Pervasive computing is the trend towards increasingly ubiquitous, connected computing devices in the environment, a trend being brought about by a convergence of advanced electronic and particularly, wireless technologies and the Internet (http://searchnetworking.techtarget.com/sDefinition/0, sid7_gci759337,00.html). Pervasive computing devices are not personal computers as we tend to think of them today, but are very tiny–even invisible–devices, either mobile or embedded in almost any type of object imaginable, including cars, tools, appliances, clothing and various consumer goods–all communicating through increasingly interconnected networks.

Sentient computing is yet another step in the general direction of ambient intelligence, Pervasive computing and Ubiquitous computing as described above. Pervasive computing is a form of ubiquitous computing which uses sensors to perceive its environment and react accordingly. A common use of sensors is to construct a world model which allows location-aware or context-aware applications to be constructed.

Also the visionaries are predicting; Wearable computer, Context-aware pervasive systems

http://en.wikipedia.org/wiki/Context-aware_pervasive_systems,Virtual reality

http://en.wikipedia.org/wiki/Virtual_reality, Human-centered computing

http://en.wikipedia.org/wiki/Human-centered_computing etc.

All of these visions point to a similar trend in that the environments become sensitive to the user and the devices in the users' surroundings start cooperating to fulfill users' needs and wishes.

Examples of this Vision

To illustrate, following are a few examples of this vision:

Intelligent Shopping Cart

Germany's Institute for Graphical Information Processing (Darmstadt), which is part of the Fraunhofer Gesellschaft, is developing the "BERNIE" shopping advisor. "It accompanies customers around the food store and warns them if the ingredients in their shopping carts do not meet their personal requirements or could even damage their health. BERNIE acquires this information by scanning the RFID tags attached to the products and matching them against the user's personal profile. Vegetarians can thus be alerted to traces of animal products and people with allergies to allergenic additives. Recovering alcoholics, diabetics and patients with high blood pressure can also be warned about potential dangers".

Figure 3.1: Intelligent Shopping Cart

Phone for Elderly

In (Adeel *et al.*, 2006) a mobile phone is described that accommodates the elderly. The phone's software "detects" the elderly by their speech patterns–slower, less energetic and trembling, and then adjusts. Possible adaptations include:

- Increasing audio volume, and amplification of high frequencies.
- Increasing ring tone volume.
- Increasing font size.
- Simplifying menu structure, probably to the detriment of features.

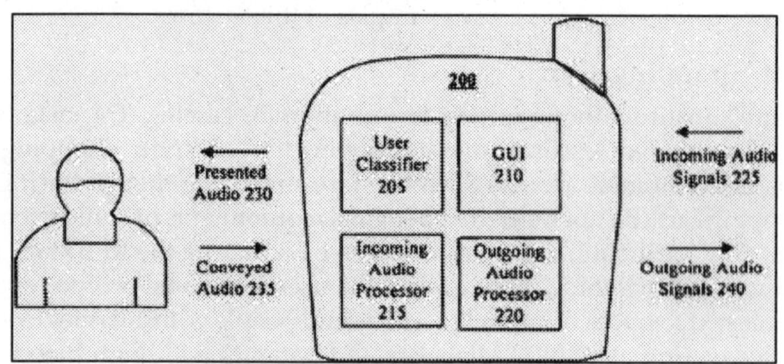

Figure 3.2: Intelligent Elderly Phone

Odour Recorder and Reproducer

The Tokyo Institute of Technology also showed an "odour recorder"(Wyszynski *et al.*, 2005). The system–the brainchild of Takamichi Nakamoto–can pick up a nearby scent and then store or transmit it so it can be reproduced on demand. An object's aroma is translated into a "digital recipe" that can be reproduced later via a selection of "original smells" used by the device.

The system could be used to improve online shopping by allowing you to sniff foods or fragrances before you buy, to add an extra dimension to virtual reality environments and even to assist doctors treating remote patients by recreating bile, blood or urine odours that might help a diagnosis. A number of companies have produced aroma generators designed to enhance computer games or TV shows, but these have been very limited in the range of smells they can produce.

Above, a few examples of ambient intelligent applications have been mentioned. It is worthwhile noting here that these applications may seem distant in time or social setup depending upon the reader's social environment. However, very different manifestations of ambient intelligence in diverse environments will become possible in the future that would be closer to the different societies of the world with their respective values and norms. One can imagine that a farmer employs ambient intelligence to constantly monitor the weather, soil composition and takes appropriate actions to improve the yield of his crops using ambient intelligence technologies in a different context than as mentioned in examples above.

Where Do we Stand Now?

Having taken a glimpse at ambient intelligence and its promise, let us now try to understand how far are we in the consumer electronics industry from reaching the goal of Ambient intelligence. In 1990s, the consumer electronics industry made the move from analogue systems to digital systems. This allowed movement of content between devices in digital formats. Early 2000's, building on digital technology, we saw start of a shift that allowed for what is called a connected consumer system where we can connect devices together to exchange content easily. Such systems have now started appearing on the market, such as a Multi-room video recorder systems that can send video data to a TV in any room in the house and allows the user to select the video that they want from their local TV screen.

Figure 3.3 is a depiction of the steps that have been taken already and what is expected in the future.

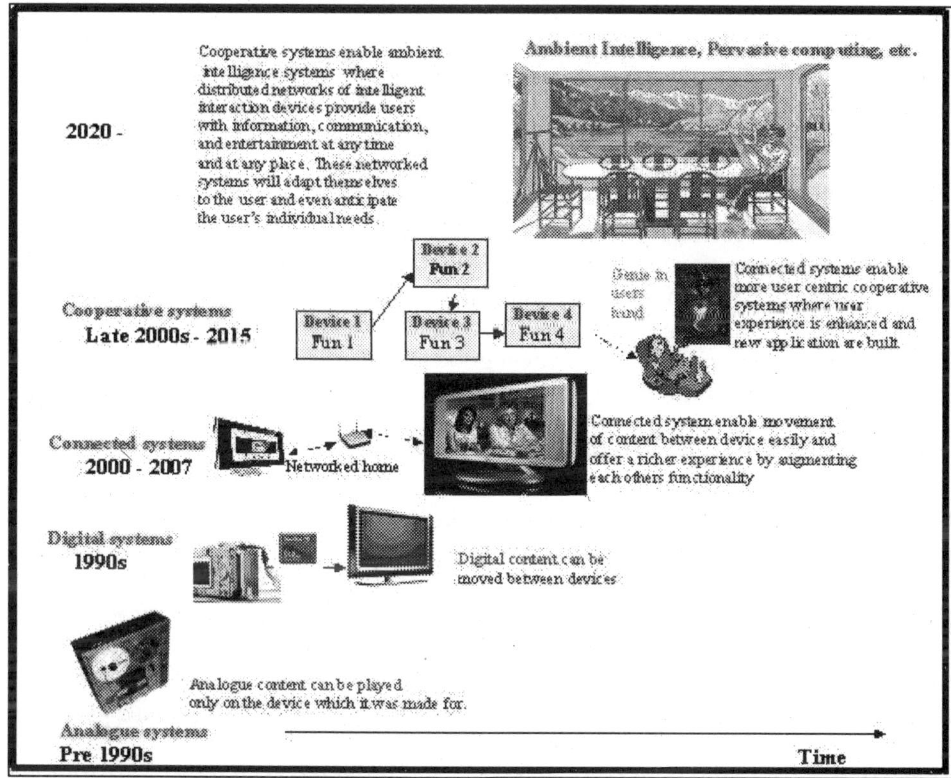

Figure 3.3: Progression in Consumer Electronic Systems

As for the next steps, we believe that the next steps towards the concept of Ambient Intelligence will lead us to cooperative systems with distributed functionality in the user's environment. In this environment, the users will have a multitude of ways or user interface methods with which they will be able to access the functionality in their environments[1].

---

1 One can think of turning on the light using a light switch when one wants to have ones environment lit but using the same lighting elements and an audio installation create a disco environment when one wants to listen to music.

We define cooperative systems as below, a definition that we have adapted from (http://www.comp.lancs.ac.uk/computing/research/cseg/projects/evaluation/coop-systems.html) as follows:

A cooperative system is a combination of devices that facilitate communication and coordination necessary to effectively work together and provide the user with the service that he desires.

In cooperative systems, the functionality is not embedded in any single device but is distributed in the user's environment in much smaller granularity than what we see today. This would then allow the user to access the capabilities of the whole network by bringing together user-required partial functionality in an ad-hoc manner. While connected systems allowed Inter-device data exchange, cooperative systems will allow Inter-device functionality usage. The user will be able to ask for a particular function and the network will arrange itself to deliver the function. The user would feel as if they have a "Genie" at their command.

Technology Issues Needing Attention

In the following, we present some issues that require attention if we are to enable Ambient Intelligent environments. It should be mentioned here that the resolution of these issues is not necessarily through nano technology development. It is usual that market forces, engineering practices and technology developments work together to resolve many of such issues. It is however good for scientists to have a notion of what these issues are as this knowledge may help guide some of their research.

Distributed Functionality and Multi-processor SW Architecture

For the Ambient intelligent systems to be perceived as unobtrusive to the users, the devices in the system will have to be hidden in the users environment e.g in furniture items. This will make it necessary to distribute the functions among a number of CPUs (Central Processing Units) in such environment. We know already that describing a system in objects and interactions between them is an art practiced by system architects and that there are no hard and fast universal rules for doing this. The problem that we currently have for single CPU systems is to identify what are the objects in a system, how do they relate to each other and how to divide them such that they can run on a CPU in different threads to perform their tasks. In the Ambient intelligence world, where we will be forced to divide the system across different CPUs, the problem will only become bigger in that we will have to identify the objects and their relationships and map them onto multiple CPUs. The issue here is how to master this complexity in our environment and if new technology can help in designing such systems.

Inter-device Streaming

With more and more distributed functionality, inter device streaming of data will become important. Devices will stream data (*e.g.* audio/video data) not only coded and compressed such as MPEG2/4, MP3, H264 etc., but also at times even decoded and uncompressed data over the network. Some of the issues that will need to be tackled in this environment will be:

- Dynamic changes in functionality used in a device such as an audio decoder which in response to changes in the environment and request from the user has to change its behaviour. For example, if an audio post processing device becomes available in the environment, then the audio decoder has to change its routing so that the post processing functionality is used.
- Definition of QoS (Quality of Service) and resource management and redundancy.
- It will become very important for the devices in the environment to become Interoperable. This will in turn mean that devices and discover each other and can collaborate to perform a function for the user. The issue will be how to build a network of devices that can negotiate resources and functionality with each other while allowing integration some new functionality that will be developed in the future and was not known when the devices forming the network were put in the network. These devices will not know about the new function and that is why it will be difficult to integrate new devices in the network.
- Another issue in cooperative systems is how to define an error, how to propagate an error and eventually how to correct an error.
- In a cooperative system, many events can occur concurrently as the event generators such as sensors etc. would be functioning autonomously and in parallel. This would mean that there is a possibility that two separate events that can affect total system behavior can occur concurrently (think of event order dependence of latencies, memory bandwidth usage, etc.). This would in turn mean that the system behavior would become unpredictable as seen from the user's point of view. While this issue can be relatively easily resolved by e.g assigning priorities in present day systems it will be very difficult to use the same trick of assigning priorities in a more dynamic cooperative system in which new devices and function enter and leave at any time. This issue remains to be resolved.

Multi-modal User Interface will be Very Important

The user interfaces (UIs) of today use a remote control and make the users walk through a menu structure. Other modalities of UI such as gesture recognition and other vision and other sensors are being investigated. Multiple sensor UI in a crude form is available in game consoles such the Wii [see www.nintando.com]. With distributed functionality of different types, multi-modal UIs will become very important. The users will not want to be bothered with menu based UIs, as is the practice today. They would expect to be able to ask for complex functions in an intuitive manner, thus requiring a sophisticated and preferably multi-modal UI based on multitude of sensors and actuators. This will require a new range of sensors and actuators such as accelerometers and camera vision etc. Nano technology could provide very non-intrusive sensors.

Another issue with sensor technologies is that they need to work together. The communication and interpretation of sensor data and results is far from understood.

For example, a camera while communicating with another camera could use a different way to describe an object than it would do when it is describing the same object to a human. Research is needed for efficient inter sensor and inter senor-actuator communication to provide responsive UI to the Ambient Intelligent world.

More Embedded Memories

Low power consideration and the fact that smaller units of local functionality that are distributed in a system will require more of fast embedded memory in our systems (*AbouGhazaleh et al.*). Also, the need is to have memory devices that are non-volatile embeddable in SOC (System on Chip) and fast enough so that they can be used as main memories for the CPU instead of the volatile SRAM memories. Some promising memory types are MRAM, FeRAM, Phase change memories etc. see for example (http://www.mraminfo.com/tags/competing_technologies).

Each of these memories have their particular issues that need to be resolved to make them fit the vision as described above.

People Identification and Environment Mapping

The Ambient intelligence systems will react to the presence of people, understand their wishes and fulfill them through collaboration among the devices constituting the system. In order for the environment to be able to do that, the Ambient intelligence system will need to be able to detect presence of a person, tell different individuals apart and in the cases when there is more than one person present in the environment, also track the movements of people. The system will also need to be able to understand the moods of different people in the environment and take appropriate actions.

In addition to knowing who is there, the system will also need to understand what the environmental conditions are and control these conditions to the liking or needs of people.

For this, many sensors and actuators will be needed such as temperature, humidity, lighting conditions etc. These sensors and actuators will have to from networks and will help control environments effectively. Some of the issues with these sensors and actuators will be in how to make them using the available nano-technologies and maybe complemented by bio technologies. How to provide power to them in such a way that they don't require battery change after some time or better, they never require a battery change.

Ad Hoc Networking

In the Ambient Intelligent world, many devices will need to work together. For this they will need to form a network, which cannot have a fixed topology as new devices and functions must have the capability of entering and leaving the network. Therefore, ad hoc networking technologies will be needed that are capable of extension but also are context aware as required in the pervasive and sentient computing visions.

Also, this network will have to be built between devices that have been built by different device makers and whose functionality is not known before hand. Some issues here would be:

- ☆ The network will consist of many devices build by many device makers. As long as the network would work flawlessly, there would be not problem but if and when there would be a fault in the network, then the issues of who provides the aftercare for devices in the network, identifies where the problem is and rectifies it will become an issue.
- ☆ Suppose a user of such a network needs to enhance some functionality in such a system. Given that most of the people would not know how to describe what exact function they need, it will be an issue as to how a consumer would know which functionality is missing from his environment and how to buy it?
- ☆ Depending on how a network is built up at a certain moment in time will depend on a number of factors such as availability of some devices, already running functions in the network, etc. For some functions guaranteed low latency on the networks will be needed. How would the system guaranteed low-latency such that the user would get the same experience every time he performs some function will be an issue.
- ☆ The devices in the network will have to be always and instantly available to the user when he needs them. From the view point of power consumption, it will be very important that the devices in the system go to a standby mode. However, even in the standby mode, devices consume considerable amount of energy (Ruth and McKenney, 2007). Thus an issue in such systems will be as to how to make devices go into a sleep state such that they do not consume any energy yet they are available instantly to the user when he needs them. One way of conserving this energy consumed in idle and standby modes would be if the devices and a domain within the connected or ambient intelligent system could be shut down but is made in such a way that it can be brought up instantly. This type of system can be made using non-volatile memories as main execution memories for devices. One example of such a connected system consisting of a media server, media renderer, control point and a proxy device has been built using MRAM (Gommans, 2005).

Waking Up Devices

If the devices do become instant-on as described in the previous section, then we will have another issue to deal with, *i.e.* how do we tell a device, which is in 'off' state, to go to 'on' state. Here is where we would like the devices to be electrically switched off but still have some way to observe the environment and react when a certain event occurs.

With MEMS (http://www.csa.com/discoveryguides/mems/overview.phpn), we could explore using mechanical means to observe the environment and trigger a device to wake up when it is needed. There is also the possibility of having sensors that can trigger the system using the observation of the environment. However, when devices are deployed in our environments, we would not expect to be changing the batteries on these devices. Some issues here are:

- The devices will have to either draw energy from the environment or be "refilled" once every so many (*e.g.* 10) years. How to make such systems such that maintenance costs are minimum and no battery changes are ever needed.
- There is a possibility to scavenge energy from our environment but the challenge here is how to make such energy scavenging devices efficient enough so that the energy spent in scavenging energy, storing it and eventually retrieving it when needed is less than the total scavenged energy.
- For existing environments and infrastructures, how to supply power to devices that do not require power in their present day implementation. For example, an intelligent door handle that reacts to presence of people replaces a present day door which normally does not require a power source for its operation. In such cases, the infrastructure has to change as in this example to lay down power leads to the door for such an intelligent handle?

Data Management

In an ambient intelligent system, there will be a lot of data that would be needed for the functioning of such a system. There will be data that the devices need to perform their required function such as what functions a device has to perform when a particular scenario plays out in its environment, there will be data about the different people in a particular environment as to who they are and what their preferences are, there will be data such as audio/video data that people use for entertainment and information. Management of such data in the system will be an issue.

How to make a data model for such a system and how to store this data in the many devices that will be in the environment such that it can be guaranteed that it can be accessed when needed and how to ensure integrity of this data. All this has of course to be done in a very energy efficient manner.

Security

Unambiguous interpretation of users wishes and feedback will be essential in ambient intelligence systems. We will need sensors to recognize people such as Imaging, biometrics using MEMS, etc.

Information from different sensing elements will need to be accumulated and decisions will have to taken based on user intentions inferred from a multitude of inputs. This information will have to be guarded against misuse. While the user interfacing is essential, the issue here will be how to ensure security of this system in a non-intrusive manner so that the user information will be secure without the user being bothered with passwords etc.

Programming Ambient Devices

It is easy to envisage that there will be multiple processing elements in ambient intelligence environments. We are used to programming our systems today using serial languages. We use threads and processes that run on a CPU and are reflection of objects and components in a system that have been identified. Problem that we face even today is that there is no exact definition of these entities and that system architecture is perceived as an art that identifies objects and components of a system. The definition of objects and components is quite arbitrary and depends on the particular philosophy used, the thinking style of the system architect and the environment surrounding the system.

In the future, multiple processor systems capable of parallel-processing will become common place as is shown in the diagram below (ITRS roadmap, 2006). In such systems the problem will become not only the identification of objects and components and which thread on a CPU to map them onto but we will have to define objects and components and also need to identify how they may run in parallel on multiple CPUs in the system.

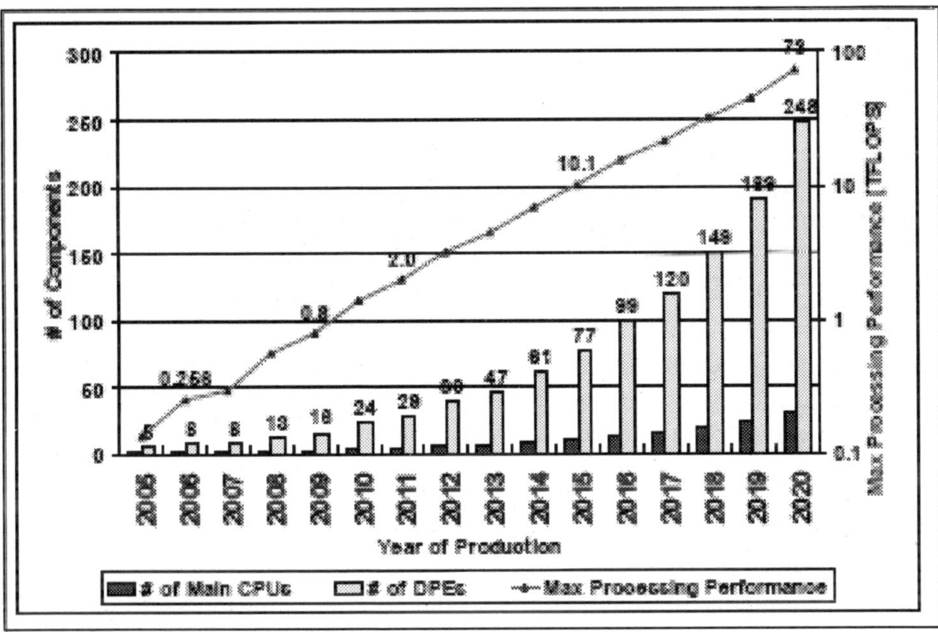

Figure 3.4: Multiple CPU Systems Expected in the Future (ITRS Roadmap, 2006)

Reliability of Ambient Device

The users today are used to a certain degree of reliability of their devices, for example a TV is a very reliable in that when it is switched on, it performs a well understood function. Today, mostly devices are built as self contained systems. In such environment, it is comparatively easier for device makers to ensure that a device is reliable. With the advent of connected systems as described presented in Figure 3.3, it is becoming clear that it is difficult to ensure reliability. Namely, it is very difficult to find the cause of a problem when a system is distributed among more devices. Thus the issue here is:

How to find out which device in a cooperative system is actually responsible for an error? To devise ways in which a system could take self correcting actions. How to do this without compromising safety, power usage, and complexity and most importantly cost in a system.

Conclusion

We have presented a set of visions for the future, which point in the same general direction, namely Ambient Intelligence. The goal is to make scientists who are busy with nanometer scale scientific developments aware of how we envisage the future consumer electronics systems. We have shown some examples that embody such visions. Finally, we have presented some issues that need resolution for these visions to become reality. We mention here that although we do not expect all these issues to be resolved by nanometer science, knowledge of these issues may help steer scientific work.

References

AbouGhazaleh, Nevine, Childers, Bruce, Moss´e, Daniel, Melhem, Rami. Energy Conservation in Memory Hierarchiesusing Power-Aware Cached-DRAM, University of Pittsburgh, Department of Computer Science 210 Bouquet St., Pittsburgh, PA, 15260, USA {nevine,childers,mosse,melhem}@cs.pitt.edu

Adeel *et al.*, 2006.Telecommunication devices that adjust audio characteristics for elderly communicators, US Patent Application 20060088154.

Gommans, Jeroen M., 2005. Instant On for Connected Devices, An Architecture for Standby/Resume and Session Transfer. Thesis Doctorate of Professional Engineering NXP/Technical University of Eindhoven the Netherlands.

http://searchnetworking.techtarget.com/sDefinition/0,,sid7_gci759337,00.html

http://en.wikipedia.org/wiki/Context-aware_pervasive_systems, Virtualreality

http://en.wikipedia.org/wiki/Virtual_reality,Human-centeredcomputing

http://en.wikipedia.org/wiki/Human-centered_computing (accessed Dec 2008)

http://www.comp.lancs.ac.uk/computing/research/cseg/projects/evaluation/coop-systems.html (accessed Dec 2008)

http://www.mraminfo.com/tags/competing_technologies (accessed Dec. 2008)

http://www.csa.com/discoveryguides/mems/overview.phpn (accessed Dec. 2008)

ITRS roadmap 2006 update

Ruth, Kurt W., McKenney, Kurtis, 2007.Energy consumption by consumer electronics in U.S. residences, Final report to the Consumer Electronics Association (CEA).

SOC Consumer Stationary Design Complexity Trends

Wyszynski, Bartosz, Yamanaka, Takao, Nakamoto, Takamichi, 2005.Recording and reproducing citrus flavors using odor recorder, 10th International Symposium on Olfaction and Electronic Noses. Volume 106, Issue 1, pp.388-393.

www.philips.com (accessed Dec. 2008)

Chapter 4

Tailored Thin Films and Nanomaterials

Kasturi Lal Chopra

Thin Film Laboratory, Indian Institute of Technology, Delhi

E-mail: choprakl@yahoo.com

ABSTRACT

Ab-initio creation of matter involves several basic processes, which include generation of atomic species as the building blocks, transport of the species through an appropriate medium, and interaction with a suitable substrate followed by nucleation and growth processes. Building blocks can be created in the form of atoms, molecules, ions, radicals, clusters, etc depending on the technique and the starting material used. Numerous techniques categorized as physical, mechanical, chemical, electrochemical, plasma, hybrids, ranging from very simple to very sophisticated ones, have been developed over the last few decades. A substrate can be chemically inert, active, reactive, or nucleation- selective. A number of deposition parameters combine to determine the structure, microstructure, topography and properties of the growing material. These parameters include the number density of the building blocks, their energy, any residual charge, and direction of incidence on the substrate, physical and chemical properties of the transport/ambient medium, physical and chemical properties of the substrate, temperature of the substrate, etc. Depending on the technique and the starting material, deposition parameters can be varied to yield Nanomatter in the form of quantum dots of different shapes, nanoclusters, nanorods, nanowires, nanotubes, sponges, very rough or atomically smooth surface thin-films, etc. Relaxed solubility conditions in the case of co-deposited species of different materials yields exotic materials in the form of man-made metastable alloys and compounds. By sequential deposition of different atomic or molecular species, Nanomatter with a periodic or periodic-aperiodic lattice of a combination of widely different starting materials can be created to yield new properties and physical phenomena and consequently new device applications. An intelligent control of deposition parameters and deposition conditions allow tailorability of some of the properties of the resulting Nanomatter. Thus, various thin-film techniques enable the creation of 0, 1, and 2 dimensional, tailorable and designer Nanomatter of considerable scientific and technological interest. This Review will highlight the developments in the broad area of the science and technology of tailored Nanomatter in the form of thin-films.

Keywords: Tailored thin films, Nanomaterials, Nanomatter, Atoms, Molecules, Nanoclusters.

Introduction

For centuries, scientists have surmised that all matter consists of atoms and molecules. Einstein was the first one to theoretically estimate, on the basis of data of diffusion of sugar in water, the size of a sugar molecule to be of nanometric (billionth of a meter) size. *NANO* is a Greek word meaning, "dwarf"; numerically it denotes "one billionth".

The creation of matter in the form of a thin film by thermal evaporation, glow discharge sputtering, and electroplating was discovered serendipitously some 150 years ago. The variable and useful electronic and optical properties of these thin film deposits led to industrial applications on an empirical basis and the field was christened as Thin Film Technologies. This discipline of science and technology established itself on scientific foundations after the availability of electron microscopes in 60's. It became clear from electron microscopy studies that thin films were nanomatter of nanometric thickness created ab-initio, by an atom (molecule)-by- atom (molecule) condensation process. Creation of such two dimensional nanomatter from building blocks of atoms / molecules / ions / monomers, manipulation of the structural arrangement of the blocks, study of their properties and utilisation of the novel and tailored properties for applications have been seriously and scientifically pursued for the last several decades. This field of science and technology occupies a pivotal role in most modern optical, electronic, opto-electronic, photovoltaic, and IC devices and surface engineering. Today, millions of kilometer square area thin-films and thin-film devices of a variety of metals, semiconductors, insulators, and polymers, and millions of tons of nanopowders of various materials are being created internationally and used in the industry for a very large range of consumer and industrial applications worth over 200 billion dollars annually. The recognition by the IC industry that smaller active electronic devices show better performance, and are faster and cheaper has been driving the research, engineering and technology development of associated nanomaterials relentlessly to the point that about ten nanometer size devices with a packing density of over 100 million devices per square cm are already routinely manufactured internationally on a very large scale.

Clearly, neither nanomaterials nor nanotechnologies are new. With dramatic developments in ultrahigh vacuum technology, computer controlled very precise monitoring and digital imaging techniques, and with the advent in 1980's and 90's of nano-analytical TEM and STM related techniques, and with the availability of reprographic and lithographic processes at nanometric scale, it has been possible to literally image atoms / molecules and to manipulate and to see the interacting atoms in the process of creating matter. The creation of matter of all varieties and its tailoring and engineering at a nanometric scale in different sizes, shapes and dimensions is now a commonplace practice. The rapid development in nano-techniques has accelerated the growth of the nano-S&T immeasurably.

Advanced nanoanalytical techniques have also shown that all living matter (plants, animals, human beings) consists of nanometric size building blocks, and molecular machines. Recent unprecedented advances in understanding and handling of nanometric size building blocks of living matter and active collaboration of physicists with life scientists have led to new interdisciplinary frontiers such as nano-bioengineering, nano-biotechnology, nano-medical science, nano-agro science, etc. As a result, a wide diversity of nano-materials and nano-technologies involving living and non-living matter of considerable interest for applications is emerging as a new S&T frontier. It needs to be pointed out that currently the adjective "nano" is being used loosely and broadly for a sub-micron size nanometric material, structure, or device, for a nanometric level process or technology, and also for any technology, which uses such nanometric material(s).

Creation of Nanomatter

Creation or synthesis of nano-materials can be achieved by "top-down", or "bottom-up" techniques. The top-down techniques are mechanical such as ball milling a material for a long time, or by beating a soft material such as silver and gold to yield nanometric size foils, a technology used by silver and gold smiths in India for centuries. The more commonly used, however, are thin-film techniques, which are particularly suited for nanodevices. Although called "bottom-up" techniques, these techniques actually involve first a "top-down" process followed by a "bottom-up" process. Depending on the process used for creating building blocks, these techniques can be categorized as follows:

- ☆ *Physical (Vapour Deposition)*: Thermal evaporation/sublimation in a controlled ambient vacuum, Laser ablation, Direct Atomic Beam writing
- ☆ *Plasma and Hybrid*: Plasma deposition using DC, RF, microwave glow discharge in various configurations, ion beam sputter deposition, Arc evaporation, Ion plating, Plasma enhanced chemical vapour deposition, Cluster beam deposition, Plasma etching
- ☆ *Electro-Chemical (Deposition)*: Nanoplating, Pulsed electrodeposition, Electrophoresis, Rapid Anodisation
- ☆ *Chemical (Vapour Deposition)*: Spray Pyrolysis, Hydrolysis, Pyrophoric, Co-precipitation, Colloidal, Sono-chemical, Laser CVD, Surface chemical-conversion, Sol-Gel, Aero and Xero –Gel, Chemical Solution/Bath deposition, Self-assembly processes, Chemical Force Microprobe
- ☆ *Lithographic/Reprographic*: Deep UV/X ray/e-Beam/Ion-beam lithography, Plasma/Ion etching, Dry lithography/reprography using chalcogenide films, Chemical and Soft lithography
- ☆ *Electrostatic*: High electric-field induced mass transfer technique using STM tip
- ☆ *Magnetic*: Magnetic Force Nanoprobe for nanomagnetic structures.

Among the various techniques, thin-film techniques are best suited for creating nanomatter with adequate controls and precision. It is important, however, to understand the atomistic processes which result in the formation of nanomatter. Sequentially, the basic processes underlying the formation of nanomatter are:

1. Creation of atomic/molecular species by breaking-up of bulk matter by a suitable process (a top-down approach)
2. Transport of a flux of species through vacuum/plasma/gas/liquid/ electrolyte, medium depending on the specie creating process
3. Adsorption, desorption,migration, interaction and coalescence of adsorbed atomic species
4. Formation of a stable cluster of critical size of atoms (critical nucleation) on a surface
5. Growth of critical nuclei both laterally and vertically, depending on the nature and properties of the flux of atomic species
6. Coalescence of nuclei (or islands), followed by sintering, and crystallization
7. Post-deposition treatments such as sintering and crystallization

Each stage in the process is affected by a number of "deposition" parameters. By manipulating and controlling some of the parameters affecting each of these stages, it is possible to tailor structural arrangement of the building blocks and thus the resultant nanomatter in various nanostructures, topographies, shapes and sizes. The various varieties of nanomatter can be categorised as follows:

Thin Film

A low dimensional (D) material, created ab-initio by condensing, one-by-one, atomic/molecular/ionic species of matter onto a substrate

Nano Material

A low dimensional material, with one or more dimensions of nanometric size, created by a thin/thick film process, or by any other material synthesis technique which yields nanosize grains in a single or composite medium.

Nano Structure

A low dimensional geometrical structure, with one or more dimension of nanometric size, obtained by a thin/thick film process and/or by micro/nano lithographic/reprographic processes.

Nano-materials are low dimensional (D) materials which, for the sake of theoretical understanding, are classified by physicists as Zero, One, or Two dimensional, depending on whether three, two, or one dimensions of the matter are sufficiently small in size (which is of course relative to the characteristic length or dimensional scale associated with the property) to constrain its properties so that the associated laws of physics are modified by the dimensional constraint and thus laws of macromatter are not necessarily applicable with confidence. Some examples of low dimensional materials and devices are:

2-D

Thin films, Thin foils/thick films, Multilayers, Superlattices, Interfaces, Surfaces of any material, Grain boundaries in materials, Quantum wells, Langmuir-Blodgett films

1-D

Whiskers, Fibres, Nano-tubes/wires/rods/bridges, Long chain molecules

0-D

Quantum Dot, Nanoparticle, Colloids, Aerogel, Nano-Porous materials (such as zeolites, suitably anodized Si, Al, Ti, etc)

Nanomaterials of different dimensions, sizes and shapes are required for different applications.

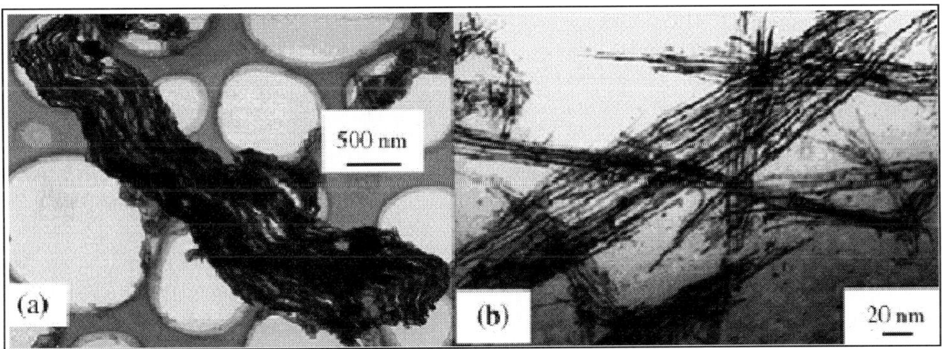

Figure 4.1: A Bright Field TEM Image of (a) Unsupported Pt nanowires on TEM grid showing agglomeration behavior after removal of silica template and (b) $P(C_6H_5)_3$ modified Pt nanowires (*J. Phys. Chem.* B, 2004, 108, 853–857)

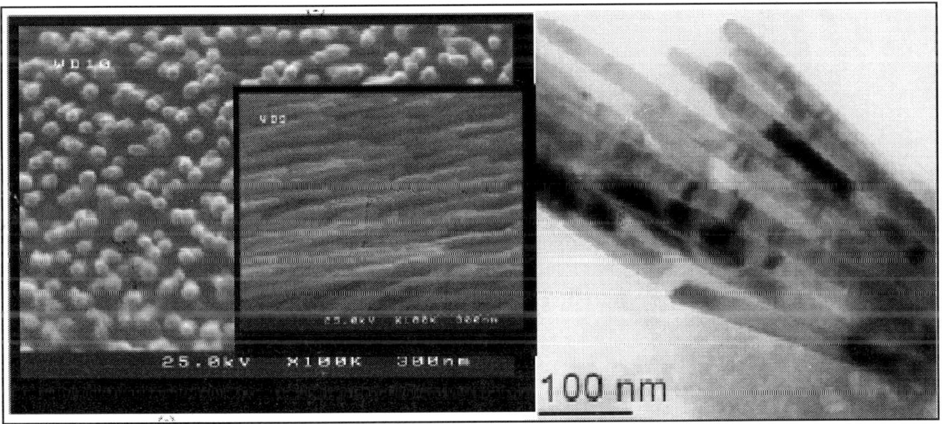

Figure 4.2: Nanorods (Bright field TEM image on right) and Cross Sectional Image of the Ge Films Deposited Obliquely on Si Subtrate at 300ºC (Ref: H. G. Chew, W.K. Choi, W. K. Chim and E. A. Fitzgerald, NUS-MIT Alliance from Google Search)

In principle, any nanomaterial can be prepared, by a suitable technique, in any shape and form such as a thin film, powder, or nanowire/nanorod. Electron micrographs (Figures 4.1–4.4) illustrate this statement nicely for four varieties of materials, *e.g.*, Pt, Ge, CdS and Y-Ba- Cu-O films.

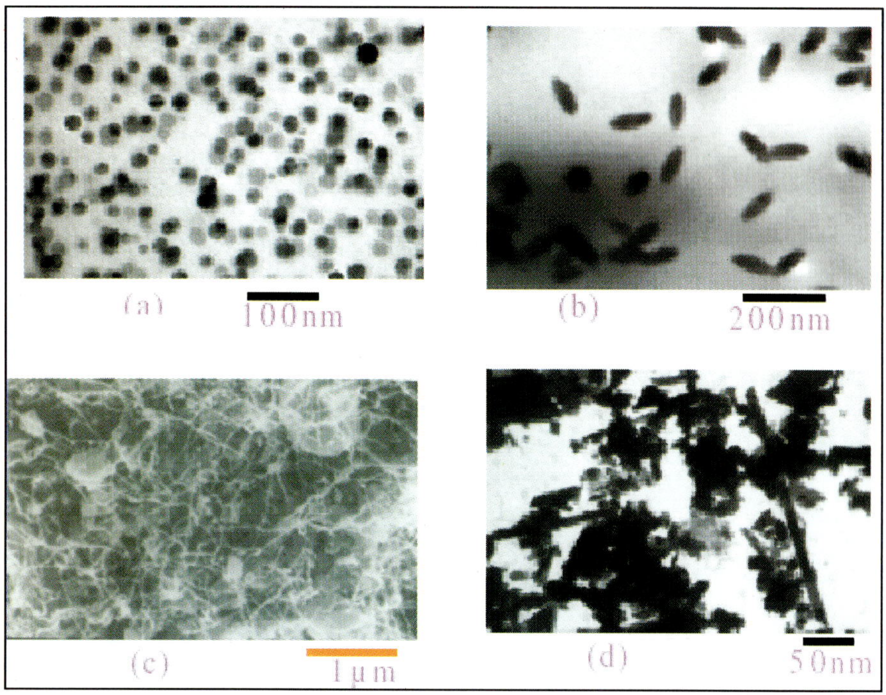

Figure 4.3: TEM Images of Spray Pyrolised CdS Films Deposited with (a) 0V; (b) 300V; (c) 700V Applied to the Spray Nozzle; (d) SEM Image of 700V film (Thin Film Lab)

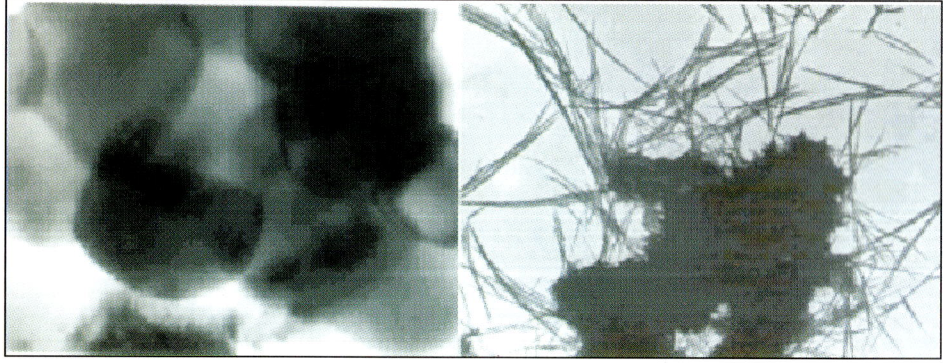

Figure 4.4: Y-Ba-Cu-O Films Obtained by Flash Pyrolysis Showing Nanostructures in the Form of Platelets, Whiskers and Tubes (Thin film Lab)

However, it must be recognized that sophisticated applications of nanomaterials in computers, communication, energy conversion, etc require specifically tailored nanomaterials of the same size, shape and orientation, and placed uniformly over a chosen substrate with a high (~ 10^{12}) packing density. This is obviously a very demanding requirement and poses a serious challenge for economic viability of nanotechnology for large scale applications. Towards this objective, a host of self-assembly techniques are being studied. Basically, the trick lies in nucleating nanomatter at designated sites which, by virtue of their chemical, metallurgical or electronic properties, have a lower nucleation barrier and thus become selective nucleation and growth centres. Some of the techniques used are: physical/chemical adsorption, sensitization of lithographically delineated sites by suitable active overlayers, and, in some cases, by energetic photon or electron beams, direct and indirect writing with atomic, electron, photon, or ion beams.

Nano-Sciences

The common feature of all nanomatter is the large surface to volume ratio. For example, for one nanometer diameter sphere, about 60 per cent atoms are on the surface and which play a dominant role in determining its physical and chemical properties. Some consequences of the dominant effect of surface atoms are:

- ☆ Surface atoms rearrange structurally
- ☆ A high density of dangling and unsatisfied bonds are created
- ☆ Surface vacancy clusters are created
- ☆ Surface energy is modified and is shape and size dependent
- ☆ Surface atomic and electronic energy states are modified
- ☆ Kinetics and dynamics of atomic and electronic interactions with each other and with ambient are enhanced and modified.
- ☆ Due to the significant role of surface atoms, the laws of physics, chemistry, metallurgy and biology and related phenomena undergo modifications, some more dramatically than others. Some properties which undergo changes are listed as follows :
  - *Thermodynamics*: Surface energy; Metastability; Solubility; Phase-Diagrams; Adsorption, Desorption, Diffusion; Melting point; Quantized Thermal Coductance; Critical phenomena; Activation Processes
  - *Physical and Chemical*: Surface activity and reactivity; Physical Density; Crystal structure; Elasticity; Hardness; Tribology
  - *Electrical*: Mean free path and Quantum size effects; Quantized energy levels; Band gap changes; Electrostatic charge and discharge effects due to modification of the Coulomb law and polarization effects for finite size charged matter
  - *Optical*: Optical constants; Optical transitions; and tectonic surface Plasmon processes

- *Magnetic*: Surface spin interactive processes; Spin tunneling, Domain structure and dynamics
- *Ferroic*: Critical phenomena of ferromagnetism, ferroelectricity, ferroelasticity, and superconductivity

Whereas qualitative changes in the properties in the Nano-Science regime can be understood to some extent, quantitative analysis is difficult and challenging, particularly in view of the fact that the changes are also size and shape dependent which may be difficult to describe analytically. For example, due to increased surface energy, one expects the melting point of a nanocluster to be lowered below the normal bulk value. Assuming a spherical shape of radius r, surface energy S, molecular volume V, and latent heat of fusion L, the depression of the melting point may be estimated by a well-known Thomson-Frenkel relation as:

$$\Delta T \sim (2S/r)\, V\, (T_B/L)$$

Measurement of the melting point of several different types of nanomatter how such behaves qualitatively. However, very little or much sharper decrease has also been reported in some cases.

Lower melting point and higher surface energy are of considerable significance in metallurgical processing since this would lead to relatively more rapid sintering at much lower temperatures.

Indeed, we have achieved in our laboratory pressure less and rapid sintering of yttrium stabilized zirconia (YSZ) nanopowder at reasonably low temperatures. This is due to the significant decrease of activation energy for sintering with decreasing size of the powder.

Similarly, by a judicious choice of deposition parameters of carbon atoms by a sputtering process and with simultaneous bombardment of the film with energetic ions during deposition, good quality diamond films deposited at room temperature have been reported in the literature.

A high density of point, line and area structural defects are inevitable when nanomatter is brought together to form a thin film or sintered to create any other shape. Vacancy concentration of 2-3 per cent and dislocations density as high as 10^{12} per cm^2 are easily created in a nanograined metal film such as copper deposited at liquid nitrogen temperature. Such films exhibit very high hardness, high electrical resistivity and enormous thermopower of 10-20 times that of bulk copper. Figure 4.5 shows an example of a giant thermopower observed by us in disordered nanograin size vapour quenched Cu-Ag films at 50:50 compositions having maximum disorder.

By depositing atomic species at an angle of incidence to the substrate, nanosized columns of the material oriented towards the atomic beam attended by a lot of porosity or low density between the columns and nanorods can be created. Such nanostructures yield a wide range of properties varying with the angle of incidence. If Ge-Se atomic beams are used, a very unusual organic polymer-like structure, first of its kind observed in any inorganic material, in the form of columns appears to be created. On irradiating such a structured film with energetic photons, electrons, or

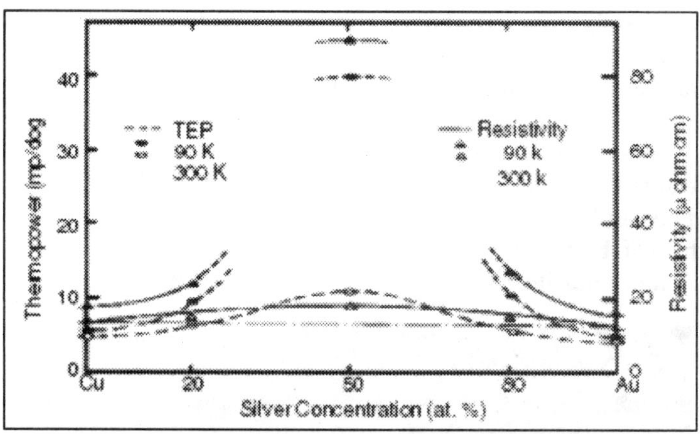

Figure 4.5: Thermopower and Resistivity of Nanocrystalline Cu-Ag Films Vapour Deposited at 90 and 300K (Thin Film Lab)

ions, the whole structure collapses to densify by as much as 40 per cent in certain cases, as shown in Figure 4.6.

This giant photocontraction effect, reported by us, has interesting applications in lithography and reprography.

Relaxation of solubility criteria and fuzzy phase diagrams of multicomponent nanomaterials are the consequence of mixing of atoms at an atomic level. An excellent example is that of incorporation of a large concentration of hydrogen and other elements in amorphous and nanocrystalline silicon, which has been utilized extensively by the 30 MW per year a-SiH thin-film solar cell industry. As another example, using a novel solution growth technique developed by us, we have successfully incorporated several metals as dopants within and between the molecular

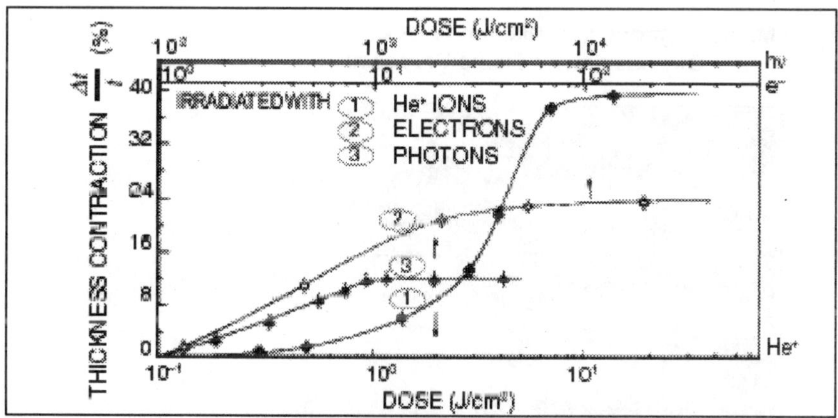

Figure 4.6: Giant Photocontraction Effect in Obliquely Deposited Nanocrystalline Ge-Se Films on Exposure to Different Energetic Radiations (Thin Film Lab)

chains of several polymers to enhance conductivity of such polymers as PVC by about 8 orders of magnitude. Since the energy difference between various possible structural arrangements of atoms of a given material is rather small, and comparable to possible additional energy component in nanomaterials due to electrostatic, magnetostatic, stress, and surface energy, observations of polymorphism and new structures are very common. The occurrence of amorphous, graphitic, diamond-like, fullerene and nanotube structures in carbon films and nanopowders is a remarkable illustration thereof. Figures 4.7 and 4.8 show electron micrographs of films of mixed diamond and diamond-like carbon (DLC), and carbon nanotubes.

Figure 4.7: Mixed Diamond and Diamond-like Carbon Films Prepared by MECVD and Oxy-aceytelene Flame Processes (Thin Film Lab)

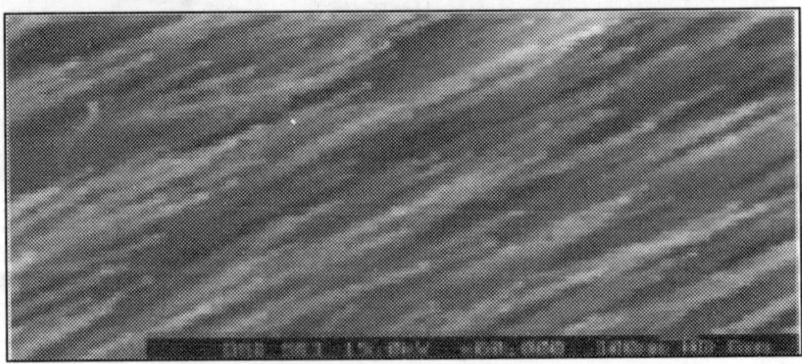

Figure 4.8: Carbon Nanotubes Prepared by a CVD Process on Ni-Fe/Cu Substrate (Courtesy Dr. Charanjeet Singh, IBM)

Some well established examples of new chemical structures in thin films are:

- ✰ Polymorphs: (*e.g.*, quartizite to spheretite and vice versa; bcc to fcc; hcp to fcc; distorted NaCl to an undistorted NaCl; etc)
- ✰ Pseudomorphs and Superstructures
- ✰ High Temperature and High Pressure Phases
- ✰ New Metastable Structures (e.g; beta Ta, and Be; fcc V, W, Mo; complex cubic Cr)
- ✰ Quasi-crystalline phases
- ✰ Multilayers, Artificial lattices and Superlattices
- ✰ Relaxed and Extended Solubility Conditions for Alloys and Compounds leading to significantly modified phase diagrams
- ✰ Mixed and Graded Chemical Compositions and Structures in ultrathin films, and multilayers.

Nanolayers

Nanolayers of different materials can be arranged and superposed in a variety of ways to yield new and artificial lattices (both periodic and non-periodic), and graded properties such as graded refractive index, band gap, work function, chemical composition, etc. An example of linearly increasing and falling index in one of the 35 periods of a superlattice (first of its kind reported in literature) consisting of alternating layers of ZnS and MgF_2 is shown in Figure 4.9. This superlattice was created by us to obtain a very high broad band reflectance filter for a high performance laser mirror. Further, excellent examples of multilayered structures designed, tailored and

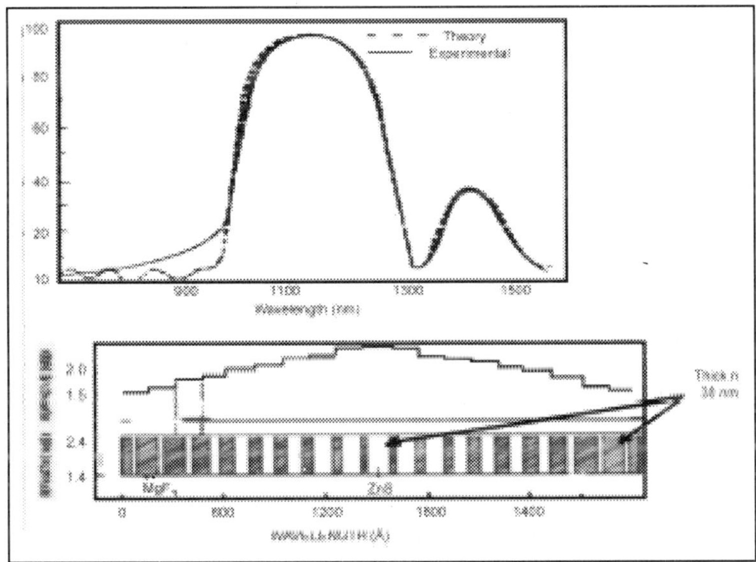

Figure 4.9: One of the 35 Graded Index Multilayer of Alternating ZnS and MgF_2 Films Designed for a High Performance, Broadband Reflection Filter for a Laser Mirror (Thin Film Lab)

engineered to optimize optical absorption, generation and collection of carriers to yield maximum efficiency of a solar cell are being hotly pursued.

By creating lattices or composites of magnetic and nonmagnetic materials, it is possible to orient and/pin spins to produce giant and colossal magnetoresistance. Spin-based changes in electronic properties constitute a new frontier, called "Spintronix", which is already being exploited commercially in information reading heads.

At sizes comparable to the corresponding Bohr radius, energy levels in a material are quantized and spatial confinement results in the increase of the energy gap, roughly proportional to the inverse of the size of the material. Such effects are now well established in a lot of semiconductors and are being effectively exploited in nanoelectronic devices based on quantum dots and quantum wells.

Some Nano- Science Phenomena and Nanotechnologies

Novel features and novel science associated with nanomatter has resulted in the discovery of many new phenomena of considerable scientific and technological interest. Some of these are:

- ☆ Giant (~ 50 μv/K) Thermoelectric power in quenched metal and alloy films
- ☆ Giant Photocontraction ($\Delta t/t \sim 40$ per cent) effect in Obliquely deposited Ge-Se films
- ☆ Giant Magnetoresistance (upto 15 per cent at RT) in magnetic/nonmagnetic multilayers
- ☆ Colossal Magnetoresistance (upto 300 per cent at LT)
- ☆ Electroresistance (12 per cent at RT); *
- ☆ Electromigration
- ☆ Graded Index Multilayers for super reflectance (99.9999 per cent) laser mirrors
- ☆ Polarized Spin Tunneling; * Tunneling Magnetoresistance
- ☆ Superhardness
- ☆ Superelasticity
- ☆ Super paramagnetism
- ☆ Super paraelectric
- ☆ Quantum confinement
- ☆ Tailored Lattices and Superlattices
- ☆ Giant Optical Absorption in Nanoparticles
- ☆ Ballistic Transport; MFP and Quantum Size Transport Effects
- ☆ Coulomb Blockade; Single Electron Transistor
- ☆ Molecular Switching

Nanomaterials continue to be utilized for a large number of applications in processes, products and devices. Some significant and large scale applications are in such areas as: Nanopowder Metallurgy; Nanocomposites; Surface Engineering; Tribological Coatings; Decorative and Optically Selective Coatings; Photonics; Quantum Electronics; Quantum Meteorology; MEMS; Nanomechanics; Solar Thermal Conversion; Thin Film Solar Cells; Ferrofluidics; Zeolites; Cosmetics; Toothpaste; Automobile Tyres; Ayurvedic Bhasam based medicines, and Gold and Silver Foils (used to decorate sweets in India).

Concluding Remarks

- ✰ With remarkable developments in sensors, monitors and controls, thin-film technologies for ab-initio creation of matter on a molecular scale can create 0, 1, and 2 dimensional nanomaterials of elements, alloy, compound, biomaterials and composites.
- ✰ By controlling deposition conditions, most nanomaterials can be created in various shapes and sizes such as smooth and rough films, wires, fibers, tubes, quantum dots and powders.
- ✰ As the length scale goes below a certain critical value, depending on the physical phenomena, a variety of new and novel physical, chemical and metallurgical phenomena are observed, giving birth to "NANOSCIENCE" in physics, chemistry, metallurgy and life sciences.
- ✰ Tailorable properties and new phenomena of low dimensional materials yield smart and efficient designer "NANOMATERIALS" for applications ranging from cosmetics to electronics, optics, mechanics, biomedical.
- ✰ Effective utilization of nanomaterials requires integration of micro with macro devices and a multi-disciplinary approach involving convergence of Physics, Chemistry, Biology, and Engineering.

Acknowledgement

I am grateful to my numerous colleagues and research students whose rich and diverse contributions in the field of thin-films and nanotechnology in our Thin Film Laboratory in IIT Delhi and Microscience Laboratory in IIT, Kharagpur form the basis of this paper.

Chapter 5

Ion Beam Applications to Next Generation Micro- and Nanoelectronics: A Brief Overview

Ahmed Shuja
International Islamic University, H-10, Islamabad, Pakistan
E-mail: shuja@solutionz247.com

ABSTRACT

This article provides a brief overview on recent advancements of ion beam applications to semiconductor process and manufacturing technology. The overview, by no means, is exhaustive but sheds some light on existing issues and status of current research related to ITRS 2007 requirements. This further includes highlights of some of the work carried out by the author in the field.

Keywords: *Micro-electronics, Ion Beam, Semiconductors, Nanoelectronics, New materials, Radiations.*

Introduction

There is a great demand of scientific work on radiation response of micro- and nanoelectronic technologies. New and emerging materials and devices are tested against the variety of radiation responses in order to examine and understand the physical phenomenon responsible for viable applications. The author, based on his independent review and analysis of semiconductor manufacturing industry, is of the opinion that today's global research on these topics may be broadly divided into six major areas:

1. Traditional radiation response with conventional radiation sources and modes
2. Study, computation and modeling of new modes of radiation applied to new materials

3. Development of novel irradiation-based device technologies
4. Study of irradiation-induced effects on semiconductor device processing
5. Single-event, displacement-damage and total fluence effects
6. Precisely positioned single ion implantation for quantum computer applications.

International Technology Roadmap of Semiconductors and Issues

Semiconductor manufacturing industry is facing stringent requirements set by back to back technology roadmaps, namely International Technology Roadmap of Semiconductors (ITRS). The process industry is having its share of research which eventually gets incorporated in diverse process routines. Some of these process routines, where radiation science is making impact and which are of much scientific interest these days, are: alternative gate dielectrics, engineered substrates, new metallization and aestivation systems, ex-conventional Si technology such as SiGe HBTs and SiGe-based Bi-CMOS technologies, ultra-narrow MOSFETs and MOSFETs with non-traditional geometry, etc. There are several research groups worldwide, which are focused on the refinement and application of a new approaches to simulate modeled and realistic individual events for subsequent radiation induced applications to micro and nanoelectronic technology. A number of contemporary applications such as SiGe Heterojunctions Bipolar Transistors, ultra-small MOSFETs, and Silicon-on-insulator are revisited with the progressively increasing level of sophisticated simulation tools (Lallement *et al.*, 2004; Ong *et al.*, 2004; Fiory *et al.*, 2001; Falepin *et al.*, 2005; Lerch *et al.*, 2005; Thompson *et al.*, 2002; Zechner *et al.*, 2002; SRIM by James F. Ziegler; SUSPRE; Eyben *et al.*, 2005; Pearton *et al.*, 1993 and de Souza *et al.*, 1996).

Other issues in such devices and other novel device concepts are also tested by the availability of first-principles calculations of conductivity/mobility in nano/sub-nano-scale MOSFETs. Defect engineering studies such as scattering from radiation-formed defects, simulation and validation of micro-dose and ultra-localized damage events in extra-small devices together with energy deposition calculations of electrically active defects in irradiated/implanted semiconductors, have also been rigorously carried out (please refer to ITRS 2007 document for more details, accessible on ITRS web site).

Radiation-induced Effects

There are varieties of temporal radiation effects, which directly or indirectly are responsible for new concepts in semiconductor manufacturing technology. Studying "Long term effects" are vitally important to understand the underlying physical phenomenon, particularly for nano-scale devices.

One must understand that these effects are no longer governed or described by spatially uniform representation of the total deposited energy. Oxide trapped charge (N_{ot}) and interface traps (N_{it}) formed by ionizing radiation at semiconductor/insulator boundaries are indeed crucial to control the current-voltage dynamics in conventional and next-generation transistor family. Electrically active defects induced by non-ionizing deposited energy directly impact the carrier mobility, lifetime, density and conductivity. Like long-term effects, transient effects cause significant changes to the device behavior under certain conditions. These effects are induced by single-events or high-fluence ionizing irradiation.

Ultra-shallow Junction Technology

Ion beam radiation or ion implantation has always been used as a mandatory tool to dope semiconductor regions. Pre-dominantly; doping and creation of Silicon-on-Insulator (SOI) wafers have been the main thrust of process and manufacturing industry. ITRS considerations are very interesting on Si and SiGe transistor family, in particular CMOS structures. One of the biggest process requirements is to develop a recipe for Ultra-Shallow Junction (USJ) technology. Dozens of research groups in academic community and R&D counterparts in global semiconductor industry are working hard to develop a readily applicable solution to create structures which contain abrupt and shallower junctions with minimal resistance and maximum dopant activation. By all means, ultra-shallow junction processing is aggressive and challenging as it encompasses a broad range of parameter controlling such as ion mass and energy, dose rate, activation mechanism of dopants, and selection of annealing schedules. There are lot of exciting reports in literature where this technology is being investigated (Lallement *et al.*, 2004; Ong *et al.*, 2004; Clarysse *et al.*, 2004; Liang *et al.*, 2005; Fiory *et al.*, 2001; Falepin *et al.*, 2005; Lerch *et al.*, 2005; Thompson *et al.*, 2002; Zechner *et al.*, 2002; SRIM by James F. Ziegler; SUSPRE; and Eyben *et al.*, 2005) keeping a range of issues and aspects in view. Author together with the commercial R&D partners are looking into the development of new activation schemes for carefully chosen implant schemes for the creation of sub 45 nm ultra-shallow junctions. Thermal budget and accurate activation measurements remain the focus and motivation of this work. Some new data on these problems will be in public domain soon.

Ion Beam Lithography

Global semiconductor industry is facing many challenges related to material selection (engineered processes) and device scaling. Process routines such as etching require various photo-masks for subsequent patterning. A mask-less solution is very much needed for next generation applications.

Dual-processing scheme for a variety of processes at the same time is also of immense importance (take an example of doping, which in case of ion implantation may be used simultaneously for patterning and etching). Feature size reduction (the main player in the nanoelectronic revolution) with the help of new probes is also of increasing importance. These wavelength associated probes are creating dramatic results in the nano-science arena. Ion beams are usually a choice when a dual process recipe is required *i.e.* doping as well as carving out the circuitry. Here carving out is purely associated with the respective wavelength of the ion species and therefore is in the nanometer resolution. Mask-less micro- and nano-beam lithography tools are not only under research utilization but rapidly coming along as big contenders of commercial electron beam and focused ion beam lithography systems. Similarly; single ions are used to precisely position single ions place and single ion tracks within a target matrix. This development may prove to be landmark for applications in quantum computing.

Implant Isolation Technology

Electrical and optical isolation technology induced by carefully developed ion implantation schedules has been used for different devices including MESFET (Eyben *et al.*, 2005; Pearton *et al.*, 1993; and D'Avanzo *et al.*, 1982), InP based HEMT(Nelson *et al.*, 1987; and Leigh *et al.*, 1985), optical wave guides(Fourre *et al.*, 1996), HBT(Fourre *et al.*, 1997; and Xiong *et al.*, 1989), PIN diodes(Pearton *et al.*, 1991; and Pearton *et al.*, 1992) and IMPATT diodes (Elliman *et al.*, 1992).

The ion species, energy and stopping mechanism, fluence, target temperature, beam current density, and post-implant time-temperature annealing cycles are important factors for isolation of III-V based material and device systems. Pearton (Pearton *et al.*, 1993) and others(Ahmed *et al.*, 2002; Boundinov *et al.*, 2001; Thompson *et al.*, 1983; Solid State Electronics by Sargunas *et al.*, 1995; Journal of Applied Physics by Sargunas *et al.*, 1995; Nuclear Instruments and Methods in Physics research by *et al.*, 1995; Almonte *et al.*, 1998; Too *et al.*, 2001; Subramaniam *et al.*, 2003; Ahmed *et al.*, 2001; Too *et al.*, 2005; Danilov *et al.*, 2002; Sargunas *et al.*, 1995; Too *et al.*, 2002; and Too *et al.*, 2003;) have reviewed the conventional carrier removal or implant isolation technology in depth, particularly for III-V devices and materials. Recently, Sealy (de Souza *et al.*, 1996) reported some major advancement in the research and applications of the topic, where some of the important contributions came from author's work (see the reference therein). The outcome of the said work is in commercial use.

Sealy's group has worked on GaAs, InGaAs and InP material/device system in detail (Ahmed *et al.*, 2005; Nucl. Instrum. Meth. Phys.by Ahmed *et al.*, 2005; Ahmed *et al.*, 2001; and Somekh *et al.*, 1974). The author continued this work on novel In-based application system for future photonic and optoelectronic devices. Our main interest lie in getting an implant isolation technology developed for InGaAsP (Too *et al.*, 2003) and InGaAsN devices.

Below is the brief review of the achieved milestones regarding InP based devices:

Table 5.1: Review of the Achieved Milestones on In-based Carrier Removal Technology for Inter- and Intra-device Isolation by Ion Implantation

Material/ Device System	*Ion Species*	*Implant Temperature (ºC)*	*Anneal Temperature (ºC)*	*Optimum Sheet Resistance (Ohms/sq)*	*Ref.*
n-InP	Proton	RT	200	2E6	26
n-InP	Proton	RT	None	4.6E4	27
n-InP	Helium	RT	380	9E6	28
p-InGaAs	Helium	RT	None	1E7	29
n-InP	Boron	RT	None	4E7	30
n-InGaAs	Neon	RT	300	1.25E5	31
n-InP	Proton	RT	None	1E6	31
n-InGaAs	Proton	RT	None	2E4	32
n-InGaAs	Iron	None	250	4E6	32

Contd...

Table 5.1–Contd...

Material/ Device System	*Ion Species*	*Implant Temperature (ºC)*	*Anneal Temperature (ºC)*	*Optimum Sheet Resistance (Ohms/sq)*	*Ref.*
p-InGaAs	Iron	None	600	7E6	33
n-InP	Helium	RT,100,200	None	~E7	34
n-InP	Iron	RT, 77K	None	~E6	35
n-InGaP	Proton	RT	500	~E9	36
n-InGaP	Helium	RT	500	~E9	36
p-InGaAsP	He, N	RT	None	~E9	37
p-InGaAs	He, N	RT	None	~E7	37
n-InP	Iron	77 K	None	5E6	38
n-InP	Iron	77K, RT, 100 C	200 C	1E7	39
n-InGaAs	Iron	77K, RT, 100 C	650 C	2.3E6	39
n-InGaAsP	Flourine	RT, 77K	500 C	~E7	40

One of our continuing works is on novel InGaAsN system for potential opto-telecommunication applications. The material system InGaAsN has recently attracted considerable attention due to its unique physical properties and possibility of creating light emitters operating in the 1.3-1.55 μm wavelength range on GaAs-based substrates(Ahmed *et al.*, 2007; Kryzhanovskaya *et al.*, 2005; Kurtz *et al.*, 2002; Ahmed *et al.*, 2001; and Ahmed *et al.*, 2005;). The strong dependence of the band gap on the nitrogen content has made this material important for a variety of opto-telecommunication applications (Ahmed *et al.*, 2007; Ahmed *et al.*, 2001; Ahmed *et al.*, 2005; Nucl. Instrum.Meth.Phys by Ahmed *et al.*, 2005; Semicond. Sci. Technol by Ahmed *et al.*, 2001; Boudinov *et al.*, 2002; Pearton *et al.*, 1990; Danilov *et al.*, 2002; Ziegler *et al.*, 1985; Hunsperger *et al.*, 1982; Moss *et al.*, 1959; Somekh *et al.*, 1974; Ahmed *et al.*, 2002; deSouza *et al.*, 1997).

III-V semiconductors are currently employed in the fabrication of optoelectronic devices and integrated circuits (IC) for high frequency operations (Ahmed *et al.*, 2007; Kryzhanovskaya *et al.*, 2005; Kurtz *et al.*, 2002; Ahmed *et al.*, 2001; Ahmed *et al.*, 2005; Nucl. Instrum.Meth.Phys by Ahmed *et al.*, 2005; Semicond. Sci. Technol by Ahmed *et al.*, 2001; Boudinov *et al.*, 2002; Pearton *et al.*, 1990; Danilov *et al.*, 2002; Ziegler *et al.*, 1985; Hunsperger *et al.*, 1982; Moss *et al.*, 1959; Somekh *et al.*, 1974; Ahmed *et al.*, 2002; and deSouza *et al.*, 1997). Electrical isolation of III–V semiconductors formed by introduction of a controlled concentration of point defects from light ion irradiation is a well-established technique (Kryzhanovskaya *et al.*, 2005; Ahmed *et al.*, 2005; Nucl. Instrum.Meth.Phys by Ahmed *et al.*, 2005; Semicond. Sci. Technol by Ahmed *et al.*, 2001; Boudinov *et al.*, 2002; Pearton *et al.*, 1990; Danilov *et al.*, 2002; Ziegler *et al.*, 1985; Hunsperger *et al.*, 1982; Moss *et al.*, 1959; Somekh *et al.*, 1974; Ahmed *et al.*, 2002; and deSouza *et al.*, 1997). Besides the efficiency of this method to obtain highly isolated layers, it has the additional advantage of maintaining the planarity of the isolated structures (Moss *et al.*, 1959; Somekh *et al.*, 1974; Ahmed *et al.*, 2002; and deSouza *et al.*,

1997). Thermal processing of the crystal tends to remove ion beam induced damage and reduce the degree of compensation, while at the same time producing diffusion of the defect centers (Kurtz *et al.*, 2002; Ahmed *et al.*, 2001; Ahmed *et al.*, 2005; Nucl. Instrum.Meth.Phys by Ahmed *et al.*, 2005; Semicond. Sci. Technol by Ahmed *et al.*, 2001; Somekh *et al.*, 1974; Ahmed *et al.*, 2002; and deSouza *et al.*, 1997). Proton implantation is also used to form infrared optical waveguides as well as other guided wave components. For such devices, annealing of the implanted regions is required in order to reduce the propagation losses to acceptable levels (Kurtz *et al.*, 2002; Ahmed *et al.*, 2001; Ahmed *et al.*, 2005; Nucl. Instrum.Meth.Phys by Ahmed *et al.*, 2005; Semicond. Sci. Technol by Ahmed *et al.*, 2001).

Therefore thermal processing due to variable implantation temperature and post-implant annealing cycles is crucial to enable electrical and optical characteristics. The results of our experiments show that the thermal stability of electrical isolation is found to depend on the original free carrier concentration and the ion dose in InGaAsN device-device matrix. A significant improvement in the thermal stability is obtained in samples irradiated to a dose we call a threshold dose, D_{th}, for higher initial sheet carrier concentrations. In these samples the isolation is maintained up to 500 °C, which seems adequate for most technological applications. Optical measurements indicate that the depths to which the optical changes extend in the as-implanted devices are greatest for liquid nitrogen implantation where no dynamic annealing occurs. Optical effects are also dose dependent. The data obtained so far shows that both electrical and optical effects are important to consider when choosing different implantation parameters for an efficient device isolation scheme. This will have ramifications for device engineers particularly in the area of optoelectronics and optical engineering. The said results together with some new findings and their possible explanation are still not available in public domain. Part of the work is under publication elsewhere.

Acknowledgements

The author would like to thanks all his academic and industrial collaborators for up to date data to prepare this review. Higher Education Commission, Pakistan for the award of the research grant (2006-2007), Radiation Physics Laboratory at CIIT, Islamabad for their cooperation and the graduate students (namely Rab Nawaz, Rehana Mustafa and Asif Bashir) to perform the laboratory work, modeling and simulation on "Ultra-shallow Junction" and "Implant Isolation Technology" projects, are also acknowledged. Further, the author also thank his ex-colleagues at IBC, UK and all industrial partners at present in USA, Canada, Malaysia, UK and EU, and Taiwan for all the support, assistance and discussion over the years.

References

Almonte, M., Yu, K. M, Haller, E. E., Ridgway, M. C., Hou, H. and Mirecki-Mullunchick, J., 1998. Proceedings of the 10th International Conference on semiconducting and insulating materials, pp. 29.

Akano, U. G., Mitchell, I. V., Shepherd, F. R., Miner, C. J., Margittai A., Svilans, M., 1996. Canadian Journal of Physics, Vol. 74, pp. S59.

Ahmed, S., Sealy, B. J., Gwilliam, R., 2002. IEEE Technical Digest: Electron Devices for Microwave and Optoelectronic Applications, pp. 18.

Ahmed, S., Gwilliam, R., Sealy, B. J., 2001. Semicond. Sci. Technol. 16, pp. 64.

Ahmed, S., Lin, J., Haq, A., Sealy, B,,2005.Nucl. Instrum. Meth. Phys. B. 240, pp. 214.

Ahmed, S., Lin, J., Haq, A., Sealy, B., Gwilliam, R., 2005. Nucl. Instrm. Meth. Phys. B 237, pp. 102.

Ahmed, S., Gwilliam, R., Sealy, B. J., 2001. Semicond. Sci. Technol. 16, pp. 28.

Ahmed, S., Knights, A. P., Gwilliam, R., Sealy, B. J., 2001. Semicond. Sci. Technol. 16 pp. 17.

Ahmed, S., Too, P., Gwilliam, R., Sealy, B. J., 2001. Appl. Phys. Lett. 79, pp. 3533.

Ahmed, S., Amirov, K., Larsson, U., Lin, J., Haq, A., Too, P., Tabatabaian, Z., 2005. Vacuum, 78, pp. 137.

Ahmed, S., Too, P., Gwilliam, R., Sealy, B. J., 2001. Applied Physics Letters, vol 79, no. 21, pp. 3533.

Ahmed, S., Nawaz, R., Syed, W. A., 2007. Applied Physics Letters, Vol. 91, no. 1, pp. 062112.

Ahmed, S., Sealy, B. J., Gwilliam, R., 2002. The 10th IEEE Intl. Sym. Elec. Dev. for Micro. Appl. (EDMO 2002), Manchester, U.K, pp. 160.

Boudinov, H., Coelho, A. V., DeSouza, J. P., 2002. J. Appl. Phys. 91, pp. 6585.

Boundinov, H., Tan, H.H., Jagadish, C., 2001. Journal of Applied Physics, Vol. 89, no. 10, pp. 5343.

Clarysse, T., Dortu, F., Vanhaeren, D., Hoflijk, I., Geenen, L., Janssens, T., Loo, R., Vandervorst, W., Pawlak, B.J., Ouzeaud, V. C. Defranoux, V.N., Faifer, M.I., 2004. Current., Materials Science and Engineering B 114–115, pp. 166–173

Carmody, C., Tan, H. H., Jagadish, C. J., 2003. Appl. Phys., Vol. 94, no. 10, pp.6616.

D'Avanzo, C., 1982. IEEE Electron Device Letters, Vol. 29, pp.105.

Danilov, I., De Souza, J. P., Boudinov, H., 2002. Journal of Applied Physics, Vol. 92, no. 8, pp.4261.

Danilov, I., De Souza, J. P., Boudinov, H. J., 2002. Appl. Phys. 92, pp. 4261.

De Souza, J. P., Danilov, I., Boudinov, H., 1997. J. Appl. Phys. 81, pp. 650.

DeSouza, J. P., Danilov, I., Boudinov, H., 1996. Appl. Phys. Lett. 68, pp. 535.

Eyben, P., Janssens, T., Vandervorst, W., 2005. Material Science and Engineering B 124-125 pp. 45-53.

Elliman, R. G., Ridgway, M. C., Jagadesh, C., Pearton, S. J., Ren, F., Lothian, J. R., Fullowman, T. R., Katz, A., Abernathy, C. R., Kopf, R. F., 1992. Journal of Applied Physics, Vol. 70, pp. 1010.

Fiory, A.T., Chawda, S. G., Madishetty, S., Mehta, V. R., Ravindra, N. M., McCoy, S. P., Lefrancois, M. E., Bourdelle, K. K., McKinley, J. M., Gossmann, H. J. L., Agarwal,

A., 2001. 9th International Conference on Advanced Thermal Processing of Semiconductors RTP 2001.

Falepin, A., Janssens, T., Severil, S., Vandervorstt, W., Felch, S.B., Parihar, V., Mayur, A., 2005. 13th IEEE International Conference on Advanced Thermal Processing of Semiconductors–RTP 2005.

Fourre, H., Schuler, O., Pesant, J. C., Leroy, A., Cappy, A., 1996. IEEE Proceedings of 8th International Conference on Indium Phodphide and Related Materials, pp.331.

Fourre, H., Peasent, J. C., Schuler, O., Cappy, A., 1997. Journal of Vacuum Sceince and Technology B, Vol. 15. no. 4, pp. 1008.

Hunsperger, R. G., 1982. Integrated Optics: Theory and Technology, Springer, Berlin.

Kurtz, S. R., Modine, N. A., Jones, E. D., Allerman, A. A., Klem, J. F., 2002. Semicond. Sci. Technol. 17, pp. 843.

Kryzhanovskaya, N. V., Egorov, A. Y., Mamutin, V. V., Polyakov, N. K., Tsatsulinikov, A. F., Musikhin, Y. G., Kovsh, A. R., Ledentsov, N. N., Ustinov, V. M., Bimberg, D., 2005. Semicond. Sci. Technol. 20, pp. 961.

Leigh, W. B., Blakemore, J. S., Koyama, R. Y., 1985. IEEE transaction on Electron devices, vol. 32, pp. 1835.

Lerch, W., Paula, S., Niess, J., McCoy, S., Selinger, T., Gelpey, J., Cristiano, F., Severac, F., Gavelle, M., Boninelli, S., Pichler, P., Bolze, D., 2005. Materials Science and Engineering B 124–125, pp. 24–31.

Liang, J. H., Sang, Y. J., Wang, C. H., Wang, T. W., Hsu, J. Y., Niu, H., Tseng, M. S., 2005 Nuclear Instruments and Methods in Physics Research B 237, pp. 312-317.

Lallement, F., Grouillet, A., Juhel, M., Reynard, J. P., Lenoble, D., Fang, Z., Walther, S., Rault, Y., Godet, L., Scheuer, J., 2004. Surface and Coating Technology 186, pp.17-20.

Moss, T. S., 1959. Optical properties of Semiconductors, Butterworths, London.

Nelson, A., Shen, Y. D., Welch, B. M., 1987. Journal of electrochemical society, Vol. 134, pp. 2549.

Ong, K.K., Pey, K.L., Lee, P.S., Wee, A.T.S., Chong, Y.F., Yeo, K.L., Wang, X.C., 2004. Materials Science and Engineering B 114–115, pp. 25–28.

Pearton, S. J., 1993. International Journal of Modern Physics B, Vol. 7, no. 28, pp. 4687.

Pearton, S. J., 1991. Nuclear Instruments and Methods in Physics Research B, Vol. 59, part 2, pp. 970.

Pearton, S. J., Ren, F., Lothian, J. R., Fullowman, T. R., Katz, A., Wisk, P. W., Abernathy, C. R., Kopf, R. F., Elliman, R. G., Ridgway, M. C., Jagadesh,C., William, J. S., 1992. Journal of Applied Physics, Vol. 71, no. 10, pp. 4949.

Pearton, S. J., 1990. Mater. Sci. Rep. 4, pp. 313.

Ridgway, M. C., Elliman, R. G., Hauser, N., 1993. Nuclear Instruments and Methods in Physics Research B, Vol: 80-1, Part 2, pp. 835.

Sealy, B. J., 2003/2004. III-Vs Review, 16, no. 9, pp. 36-38.

Sargunas, V., Thompson, D. A., Simmons, J. G., 1995. Nuclear Instruments and Methods in Physics research B, Vol. 106, pp. 294.

Sargunas, V., Thompson D. A., Simmons, J. G., 1995. Solid State Electronics, Vol. 38, no. 1, pp. 75.

Sargunas, V., Thompson, D. A., Simmons, J. G., 1995. Journal of Applied Physics, Vol. 77, no. 11, pp. 5580.

Surrey University Sputter Profile Resolution from Energy (SUSPRE), Surrey Ion Beam Centre supported by Engineering and Physical Sciences Research Council (EPSRC)

Sargunas, V., Thompson, D. A., Simmons, J. G., 1995. Journal of Applied Physics, Vol. 77, no. 11, pp.5580.

Subramaniam, S.C., Rezazadeh, A.A., Too, P., Ahmed, S., Sealy, B. J., Gwilliam, R. 12-16th May, 2003.15th Indium Phosphide and Related Materials Conference, Santa Barbara, USA.

Somekh, S., Garmire, E., Yariv, A., Garvin, H. L., Hunsperger, R. G., 1974. Appl. Opt. 13 pp. 327.

Thompson, P. E., Bennett, J., 2002. Material Science and Engineering B89, pp.211-215.

Thompson, P. E., Binari, S. C., Dietrich, H. B., 1983. Solid State Electronics, Vol. 26, no. 8, pp. 805.

Too, P., Ahmed, S., Gwilliam, R., Sealy. B. J., 2001.International Symposium on Electron Devices for Microwave and Optoelectronic Applications, Vienna, pp.125-130.

Too, P., Ahmed, S., Gwilliam, R., Sealy, B.J., 2005. Nuclear Instruments and Methods in Physics research B, Vol. 240, pp. 178.

Too, P., Ahmed, S., Jeynes, C., Sealy, B. J., Gwilliam, R., 2002. Electronic Letters, Vol. 38, no. 20, Pg.1225-1226.

Too, P., Ahmed, S., Jakiela, R., Kozanecki, A., Barcz, A., Sealy, B. J., Gwilliam, R., 2003. The 11th IEEE International Symposium on Electron Devices for Microwave and Optoelectronic Applications (EDMO 2003), Orlando, USA, 17-18th November, pp.18-23.

Xiong, F., Tombrello, T. A., 1989. Nuclear Instruments and Methods in Physics Research B, Vol. 40, pp. 526.

Ziegler, J. F., Biersack, J. P., Littmark, U., 1985. The Stopping and Range of Ions in Solids, Pergamon, Oxford.

Ziegler, James F. Stopping and Range of Ions in Matter (SRIM) by, SRIM.org, Annapolis, MD 21402, USA.

Zechner, C., Matveev, D., Erlebach, A., Simeonov, S., Menialenko, V., Mickevicius, R., Foad, M., Al-Bayati, A., Lebedev, A., Posselt, M., 2002. Nuclear Instruments and Methods in Physics Research B 186, pp. 303-308.

Chapter 6

Growth and Characterization of Single Crystalline Cubic SiC on Porous Si Using Low Pressure Chemical Vapour Deposition Technique

M. Asghar[1]*, Payam Shoghi[2], Hadia Noor[1], and M.A. Hasan[2]

[1]Semiconductor Division, The Islamia University of Bahawalpur, Pakistan

[2]Department of Electrical and Computer Engineering and the Centre for Optoelectronic and Optical Communication, University of North Carolina Charlotte, NC 28223

*E-mail: *asghar_hashmi@yahoo.com, prof_asghar@iub.edu.pk*

ABSTRACT

Single crystalline 3C-SiC layers were grown using a single gas source trimethylsilane on a nanoscale porous Si seed in a low pressure CVD reactor fitted with an RF heating stage capable of heating the samples up to 1250°C. The single-crystalline porous skeleton of Si has been used as an elastic template to accommodate the large lattice mismatch between the Si substrate and the overlayer. Porous Si layers were made by anodizing p-type Si(100) wafers in a mixture of hydrofluoric acid and ethanol. FTIR spectra showed a strong peak at 800 cm-1, which corresponded to the Si-C vibrational mode. Large area XRD showed a single peak at 2è value ~ 42o with FWHM of ~0.57o originating from 3C-SiC (200) orientation. However, depending on the growth conditions, other orientations such as SiC (111) and SiC(110) were also observed in some samples. Cross-sectional SEM image of the SiC/Si interface showed ~ 800nm thick SiC layer but loosely bonded on Si and embedded with large number of voids, thereby optimizing for the separation of SiC from the Si for full SiC wafers production.

The structure and properties of the SiC layer are correlated to the preparation condition of the porous layer, growth temperature, gas flux, and hydrogen treatment prior to growth.

Keywords: *3C-SiC, LPCVD, Trimethylsilane, Defects, XRD, FTIR, SEM.*

Introduction

Silicon is the material dominating the electronics industry today. However, attractive properties of silicon carbide (SiC) have outclassed Si in many applications and it is considered now as the material of choice for future electronic devices operating under harsh environment of temperature and radiation (Chung and Kim, 2007; Mehregany and Zorman, 1999; Sarro and Sen, 2000). Due to its wide bandgap, heat and other external influences do not readily disrupt the performance of SiC microelectronics (Chen *et al.*, 2003). Therefore, SiC devices can operate at higher temperatures and higher radiation levels than silicon- and gallium arsenide-based devices (Reyes *et al.*, 2006; Gupta *et al.*, 2004; Itoh *et al.*, 1996). SiC can also operate at higher power levels, and emit blue light (Fan *et al.*, 2006). Some other properties of SiC include a high electron velocity (which means that SiC devices can operate at high frequencies), a high thermal conductivity (which means that SiC devices can easily dissipate excess heat), and a high breakdown electric field (which means that SiC devices can operate at high voltage levels ~ 5000V) (Morkoç *et al.*, 1994; Förster *et al.*, 2005).

Most of these developments have occurred in the hexagonal 4H- and 6H-polytype of SiC substrates, yet, the high cost of the substrate (SiC) limits large-scale commercialization of this material in the industry (Semmelroth *et al.*, 2004).In the meantime, the growth of 3C-SiC on cheap Si substrate is one of the most promising technologies for bringing the benefits of conventional SiC to consumers (Nagasawa and Yamaguchi, 1992; Steckl *et al.*, 1993; Seo *et al.*, 1997). Nevertheless, due to large differences in lattice constant (20 per cent) and thermal expansion coefficient (8 per cent) of the two materials, heteroepitaxy of SiC using Si as a substrate always results in growth of 3C-SiC with a high density of crystallographic structural defects such as stacking faults, microtwins, and inversion domain boundaries(Gupta *et al.*, 2007; Asghar *et al.*, 2007).Recently, porous Si has been reported to provide nanoscale seeds for the growth of reasonably pure SiC but still the quantization of the related growth conditions is an open question(Madapura *et al.*, 1999; Steckl and Li,1992).

In this paper, we report growth of SiC-on-porous Si by low pressure CVD technique for onward technological revolution especially in power electronics. Single crystalline 3C-SiC layers were grown in a low pressure CVD reactor fitted with an RF heating stage capable of heating the samples up to 1250°C. The bonding of Si-C was confirmed by FTIR while the crystal structure was examined by SEM and XRD.

Experimental Details

Single crystalline 3C-SiC layers were grown using a single gas source on a nanoscale porous boron-doped Si seed with NA ~ 1017 cm^{-3}. Single-crystalline porous skeleton of Si was used as an elastic template to accommodate the large lattice mismatch between the Si substrate and the overlayer. This method is economical, environmental friendly, and utilizes a non-toxic and non-pyrophoric gas; trimethylsilane. Porous Si layers were made by anodizing p-type Si (100) wafers at 5 mA DC current for 10 min, in a mixture of hydrofluoric acid and ethanol filled in Teflon cylinder.

Detailed preparation scheme is illustrated in the following (c.f. Figure 6.1): porous silicon wafer is placed on the cylindrically shaped graphite susceptor (dia= 50 mm, height = 100 mm) in the main chamber. Chamber is closed and rotary pump is switched on. At chamber pressure about 10^{-3} Torr, the turbo molecular pump (TMP), is connected to the chamber by opening the specific valve, in this way the pressure inside the chamber approaches to 10^{-5} Torr. During this period, the graphite susceptor is heated by an RF-coil operated by DC-RF power supply, and the temperature reaches 1160 ºC as measured by infra-red thermometer. Next, precursor gas, trimethylsilane $SiH_4(CH_3)_3$ is allowed to flow into the chamber, especially in the vicinity of graphite cylinder. Due to hot environment, C-H and Si-H bonds are broken into Si and C ions, which in turn, interact with nanoscale seeds on porous Si substrate: a chemical reaction takes place on the surface; resultantly SiC film is grown on Si substrate while the desorbed products are transported away from the surface. Chemical reaction can be described by following equation:

$$2SiH_4(CH_3)_3 + Si \rightarrow 3CH_4\ (gas) + 3Si + 3C + 3H_2\ (gas)$$

The flow of the gas is maintained at 0.1-0.2 sccm (standard cubic centimeter per minute). The process was kept on for about 4 to 6 hours to have hundreds of nanometer film of 3C-SiC on Si. A schematic diagram of the LP-CVD is shown in Figure 6.2.

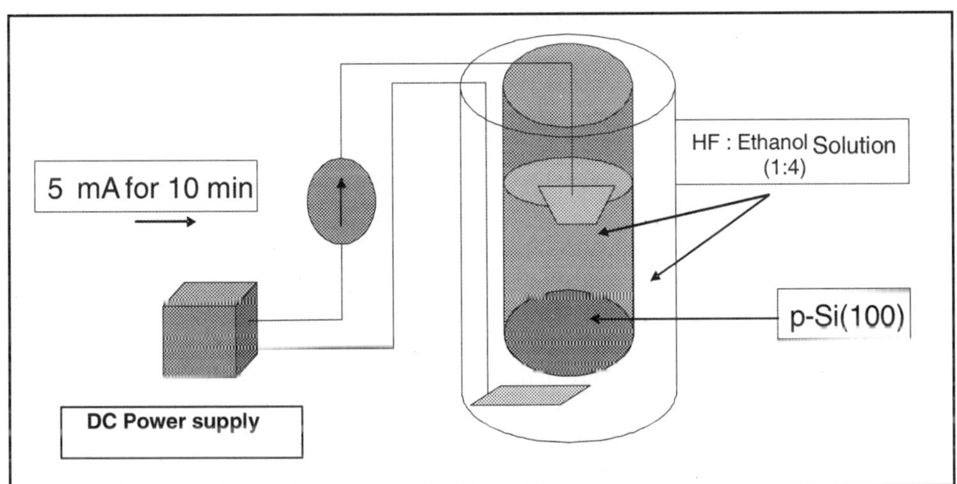

Figure 6.1: A Teflon Cell Used for Preparation of Porous Si.

For cleaning a solution of de-ionized water (100ml) and sulphuric acid (150ml) was used. Electrochemical: A solution of HF 49 per cent (150ml) and ethanol (600 ml) was employed in electrochemical reactions.

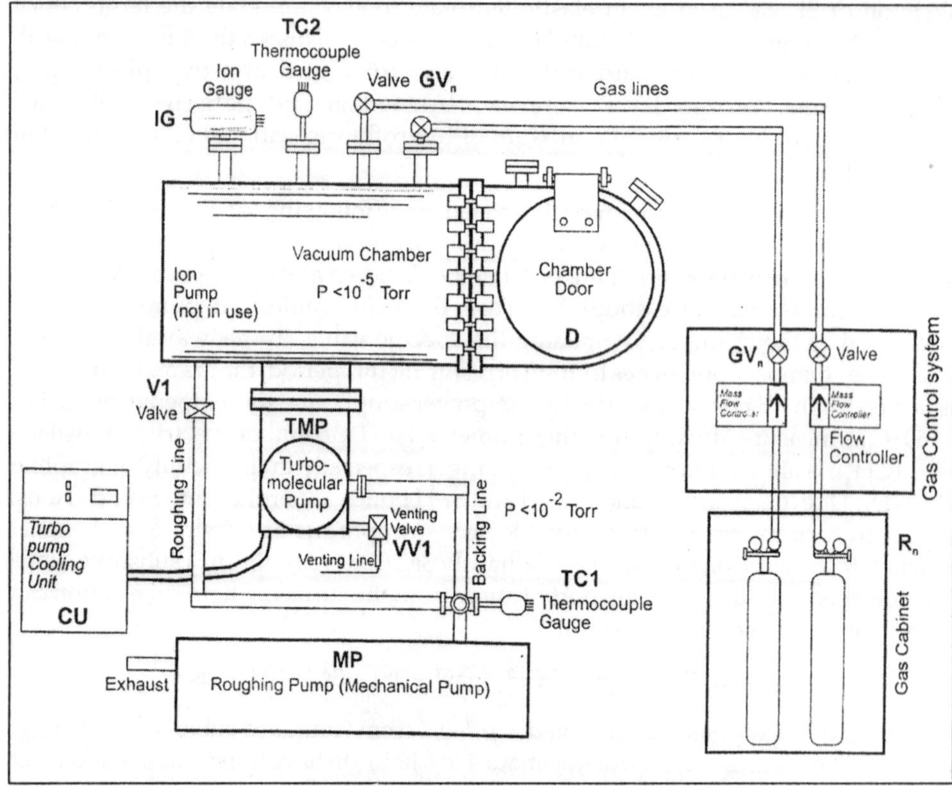

Figure 6.2: A Schematic Diagram of the Low Pressure Chemical Vapor Deposition System

Fourier transform infra-red spectroscopy (FTIR) confirms Si-C bonding while crystal structure and thickness of 3C-SiC have been examined by X-ray diffraction and cross sectional (SEM), respectively.

Results and Discussion

Figure 6.3 shows an X-ray diffraction pattern recorded with a SiC layer grown at 1160 °C on porous Si layer prepared using anodization current of 5 mA. The Figure shows strong peaks at 2θ: 42° and 91° attributed to 3C-SiC (200) and Si(400) (the substrate), respectively; the pattern indicates growth of a relaxed single crystalline SiC layer. Our observed values are in agreement with those reported by Förster *et al.*, and Gupta *et al.*. However, depending on the growth conditions, other orientations such as SiC (111) and SiC (110) were also observed in our samples (c.f. Figure 6.3). Using Scherrer formula (Mendoza-Galvan *et al.*, 2006), crystalline size of the grown 3C-SiC layer was found to be 17nm, which was understandably small due to strains and/or larger lattice mismatch (Chen *et al.*, 2007). The structure and the properties of the SiC layer are correlated to the preparation condition of the porous layer, growth temperature, gas flux, and hydrogen treatment prior to growth (Madapura *et al.*,

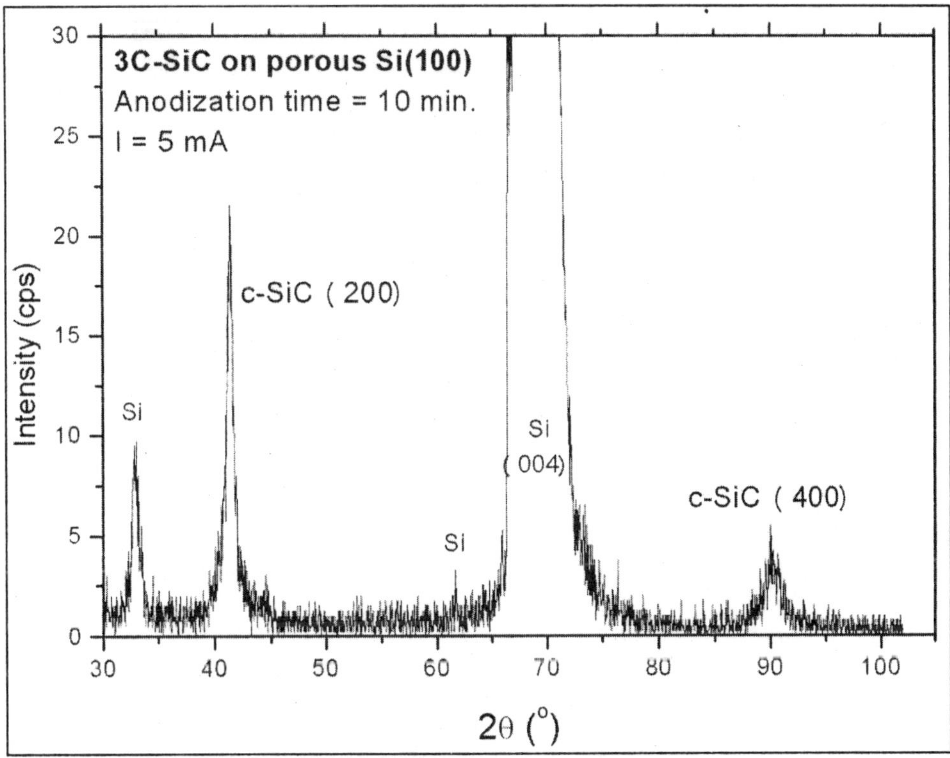

Figure 6.3: An X-ray Diffraction Spectrum of 3C-SiC (100) on Porous P-type Silicon Substrate Grown at 1160 ºC

1999). Figure 6.4 demonstrates FTIR spectrum of the layer showing a dominant peak at frequency 800 cm^{-1} which corresponds to the Si-C vibrational mode (Perevedentseva *et al.*, 2007; Chen *et al.*, 2000). In addition, the FTIR spectrum also yields two additional peaks associated with Si-Si (substrate) and C-O at 611 cm^{-1} and 1105 cm^{-1} respectively. This observation is also consistent with our XRD results. Figure 6.5 shows a cross-sectional scanning electron microscope (SEM) image of SiC layer grown on Si where we can see about 800nm thick SiC layer being loosely bonded on Si. This observation will allow the separation of SiC from the Si for mass production of low-cost and large-area SiC wafers. The reason for SiC layer to be loosely bonded on Si is attributed to the presence of voids at SiC/Si interface (Madapura *et al.*, 1999); consequently the SiC layer is optimized to be separated from Si substrate for mass growth of SiC wafers on SiC substrate.

Conclusion

Single crystalline 3C-SiC layers were grown using trimethylsilane as a single gas source on a nanoscale porous Si seed in a low pressure CVD reactor fitted with an RF heating stage capable of heating the samples up to 1250 °C. Bonding of SiC was confirmed by Fourier transform infra-red spectroscopy (FTIR) while the crystal

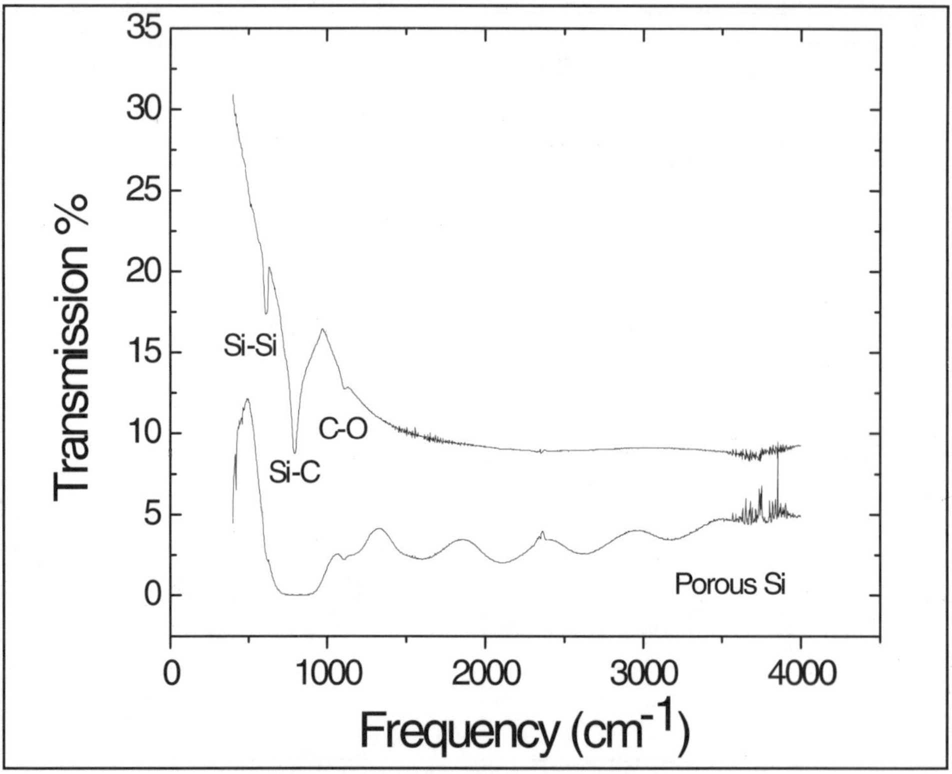

Figure 6.4: A Typical FTIR Transmission Spectrum of 3C-SiC Film Grown on Porous Si(100) Substrate, $T_{Substrate}$ =1160 ºC

Figure 6.5: A Scanning Electron Microscope Image of SiC(100) Loosely Bonded to Si Substrate. Further optimization would allow the separation of SiC from the Si for growth of SiC wafers.

structure was examined by scanning electron microscopy (SEM) and X-ray diffraction. The salient features are listed as under:

FTIR showed a strong peak at 800 cm^{-1}, which corresponds to the Si-C vibrational mode.

Large area X-ray diffraction showed a single peak originating from 3C-SiC (100). However, depending on the growth conditions, other orientations such as SiC (111) and SiC(110) were also observed in some samples.

A scanning electron microscope image of SiC (100) showed a loosely bonded SiC layer on Si substrate. The loosebonding between SiC and Si is attributed to the presence of voids at SiC/Si interface.

The structure and the properties of the SiC layer are correlated to the preparation condition of the porous layer, growth temperature, gas flux, and hydrogen treatment prior to growth.

Acknowledgment

The authors are thankful to Higher Education Commission of Pakistan and Charlotte Research Institute, UNCC for providing financial assistance under POCR program and research facilities, respectively. The authors are also grateful to Prof. Dr. Yasin Akhtar Raja for making the research activity convenient.

References

Asghar, M., Hussain, I., Noor, H.S., Iqbal, F., Wahab, Q., Bhatti, A.S., 2007.J. Appl. Phys. 101, pp.073706.

Chen, ChengChun, Yu, BenHai, Liu, JiangFeng, Dai, QiRun, Zhu, YunFeng, 2007.Mater. Lett. 61, pp.2961.

Chen, J., Steckl, A.J., Loboda, M.J., 2000.Mater. Sci. Forum. 338-342, pp.273.

Chung, G.-S. and Kim, K.-S., 2007.Bull. Korean Chem. Soc. 28, pp.533.

Chen, X.D., Fung, S., Ling, C.C., Beling, C.D., Gong, M., 2003.J. Appl. Phys. 94, pp.3004.

Fan, J.Y., Wu, X.L., Li, H.X., Liu, H.W., Siu, G.G., Chu, Paul K., 2006.Appl. Phys. Lett. 88, pp.041909.

Förster, Ch., Cimalla, V., Brückner, K., Hein, M., Pezoldt, J., Amabacher, O., 2005. Materials Science and Engineering C. 25, pp.804.

Gupta, A., Paramanik, D., Varma, S., Jacob, C., 2004. Bull. Mater. Sci. 27, pp.445.

Gupta, A., Sengupta, J., Jacob, C., 2007. Thin Solid Films, xx,xxx-xxx (article in press).

Itoh, A., Kimoto, T., Matsunami, H., 1996. IEEE Electron Device Lett. 17, pp.139.

Mehregany, M., Zorman, C.A., 1999. Thin Solid Films. 355, pp.518.

Morkoç, H., Strite, S., Gao, G.B., Lin, M.E., Sverdlov, B., Burns, M., 1994. J. Appl. Phys. 76, pp.1363.

Madapura, S., Steckl, A.J., Loboda, M., 1999. Journal of the Electrochemical Society. 146, pp.1197.

Mendoza-Galvan, A., Trejo-Cruz, C., Lee, J., Bhattacharyya, D., Metson, J., Evans, P.J., Pal, U., 2006. J. Appl. Phys. 99, pp.014306.

Nagasawa, H., Yamaguchi, Y., 1992. Springer Proceeding in Physics, Springer-Verlag, Berlin 71, pp. 40.

Perevedentseva, E., Tung, F.-K, Chung, P.-H., Chou, P.-W., Cheng, C.-L., 2007. Thin Solid Films. 515, pp. 5259.

Reyes, M., Waits, M., Harvey, S., Shishkin, Y., Geil, B., Wolan, J.T., Saddow, S.E., 2006.Mater. Sci. Forum. 527-529, pp.307.

Sarro, Sen, P.M., 2000. Actuators A. 82, pp.210.

Semmelroth, K., Krieger, M., Pensl, G., Nagasawa, H., Puschl, R., Hundhausen, M., Ley, L., Nerding, M., Strunk, H.P., 2004. Mater. Sci. Forum. 457-460, pp.15.

Steckl, A.J., Yuan, C., Li, J.P., Loboda, M.J., 1993.Appl. Phys. Lett. 63, pp.3347.

Steckl, A.J., Li, J.P., 1992.IEEE Trans. Electron Devices. ED-39, pp. 64.

Seo, Y.H., Nahm, K.S., Suh, E.K., Lee, H.J., Huang, Y.G., Vac. J., 1997.Sci. Technol. A15, pp.2226.

Chapter 7

Recent Advances in Structural Characterization of Single Crystals by High Resolution X-ray Diffraction

Krishan Lal
National Physical Laboratory, New Delhi – 110 012, India
E-mail: klal@mail.nplindia.crnct.in

ABSTRACT

Single crystals are building blocks of several key technology areas including microelectronics. These are required to conform to very stringent specifications of properties and characteristics like elemental composition and real structure. Other parameters like crystallographic orientation are also of considerable importance. Therefore it is vital to have capabilities to evaluate starting materials and also for monitoring changes due to processing, like thin film deposition, ion implantation and metallizations. High-resolution X-ray diffraction techniques have been extensively employed for non-destructive evaluation of structural quality of single crystals, thin films, interfaces and devices. The demand on accuracies for determination of structural parameters like biaxial stress is continuously increasing and therefore, posing new challenges for scientists. In author's laboratory, a series of six multicrystal X-ray diffractometers have been developed for: (i) high resolution X-ray diffractometry, topography and diffuse X-ray scattering measurements; (ii) study of electric field induced defects in crystals and microelectronics devices; (iii) measurement of biaxial stress in thin film-single crystal substrate systems; (iv) accurate determination of crystallographic orientation of crystals; (v) accurate lattice parameter measurements (1-10 ppm); (vi) measurement of lattice and orientational mismatches between epitaxial films and their substrates; (vii) anomalous transmission and absorption of X-rays through crystals and (viii) accurate measurement of absorption coefficient of single crystals for X-rays. The most sophisticated of these systems is a *Five Crystal*

X-ray Diffractometer. This diffractometer employs three plane silicon monochromators, set in (+,-, +) geometry. These incombination with a long collimator enable achieving a nearly parallel exploring X-ray beam with wavelength spread negligible in comparison to the intrinsic spread of the characteristic radiation. Diffraction curves half widths for 400 reflection of silicon crystals when recorded with this system are only ~1.7 arc sec in comparison to the theoretical value of 1.45 arc sec (Mo $K\alpha_1$ radiation; Bragg geometry). In this presentation results on studies of defects in single crystals, (as grown as well as those subjected to processing) will be presented. Biaxial axial stress in thin film-single crystal substrates has been investigated in several systems with silicon and gallium arsenide substrates.

Keywords: *Crystal, Resolution, Technology, X-ray.*

Introduction

Single crystals are being extensively employed in fabrication of a wide variety of devices including those used in microelectronics and nano technology. In the field of microelectronics, the last few decades have seen great success in reduction in feature size going down to a few tens of nano meters and increase in device density (Bullis and Huff, 1993; Claeys and Deferm, 1996). The processes for fabrication have grown in complexity and the levels of integration and stringency in specifications have gone up. The starting single crystals are required to conform to strict limits in terms of elemental purity, structural perfection, physical properties, and parameters like crystallographic orientation (Laudise, 1972; Lal, 1991). As these wafers are subjected to processing like oxidation, lithography, implantation, etching and annealing etc., the changes in microstructure and impurity distribution are required to be monitored closely. Information about all types of defects like boundaries, dislocations and point defects and their clusters is required with high level of reliability. High-resolution X-ray diffraction techniques have been successfully employed for evaluation of structural quality of single crystals, thin films, interfaces and solid surfaces (Lang, 1958; Bowen and Tanner, 1998; Kato, 1992; Lal, 1993). The author and his co-workers have developed high-resolution X-ray diffraction techniques and multi crystals X-ray diffractometers to meet these requirements. The most sophisticated of the diffractometers, is a five crystal system which has state-of-the-art level resolution (Lal, 1993; Lal and Singh, 1977; Lal and Bhagavannarayana, 1989; Lal *et al.*, 1990; Lal, 1998). This system enables one to carry out a wide variety of experiments at very high resolution. It can be used for: (*i*) high resolution X-ray diffractometry, topography and diffuse X-ray scattering measurements; (*ii*) study of electric field induced defects in crystals and microelectronics devices; (*iii*) measurement of biaxial stress in thin film-single crystal substrate systems; (*iv*) accurate determination of crystallographic orientation of crystals; (*v*) accurate lattice parameter measurements (1-10 ppm); (*vi*) measurement of lattice and orientational mismatches between epitaxial films and their substrates; (*vii*) anomalous transmission and absorption of X-rays through crystals and (*viii*) accurate measurement of absorption coefficient of single crystals of X-rays.

Several investigations in the domain of fundamental research as well as in applied fields have been carried out by employing high-resolution X-ray diffraction. In this paper, some important results that are relevant to the needs of semiconductor industry are presented. Essential theoretical background and basics of the experimental techniques are also described.

High Resolution X-Ray Diffraction Techniques

Theoretical Background of the High-Resolution X-Ray Diffraction Techniques

Basically two theoretical approaches are employed to understand diffraction of X-ray from crystals. The kinematical theory is widely employed when the crystal sizes are small, like powders and/or when the degree of perfection of crystals is low. The more rigorous and elaborate dynamical theory is employed while dealing with nearly perfect crystals of large sizes (Zachariasen, 1945; Laue, 1960; Pinsker, 1978; Batterman and Cole, 1964; Authier, 2001). Therefore, in the case of crystals employed for device fabrication like the semiconductor crystals, the dynamical theory of X-ray diffraction has to be employed for obvious reasons. In the following, we shall describe the basic features of the dynamical theory particularly of relevance to the topic of this paper.

The basic formulation of dynamical theory starts by considering propagation of X-rays as electro-magnetic waves through a crystal, which can be treated as a material with periodically varying dielectric constant (Zachariasen, 1945; Laue, 1960; Pinsker, 1978; Batterman and Cole, 1964; Authier, 2001). The periodicity is that of the crystal lattice itself. When the X-ray beam propagates through crystals, each sheet of atoms (lattice plane) becomes a source of scattering due to interaction of the electric vector of the X-ray beam with the electrons localized on atomic positions.

As the beam proceeds inside the crystal, there are a number of scattered beams in addition to the original beam whose intensity has been diminished due to interaction with atomic layers as a result of scattering and absorption. Any atom in the interior of the crystal will not only receive the depleted incident beam but also scattered waves from all the other atoms that have been irradiated by the exploring beam. To understand the features of the diffracted beam, one has to consider interaction between the scattered and the exploring X-ray beams as well as that between the different scattered waves. The most essential steps needed for calculation of diffracted X-ray intensity in the case of reflection geometry (Bragg Case) as well as in the transmission configuration (Laue Case) are described in the following.

The propagation of X-rays can be understood in terms of Maxwell's equation inside the crystals. Maxwell's equation can be written as [13]:

$$\nabla \times \mathbf{E} = -\partial \mathbf{B}/\partial t = -\mu_0 \partial \mathbf{H}/\partial t \quad [1]$$

$$\nabla \times \mathbf{H} = \partial \mathbf{D}/\partial t = \varepsilon_0 \partial (\kappa \mathbf{E})/\partial t \quad [2]$$

Here, **E**, **D**, **B** and **H** are the four vector fields representing the electric vector, the displacement vector, magnetic vector and the magnetic induction vector, respectively.

The conductivity σ of the crystal has been taken as zero at the very high X-ray frequencies. The crystal is assumed as non-magnetic. For convenience of treatment, it is assumed that the crystal is oriented in such a manner that it is quite close to a diffraction maximum, so that:

$$\mathbf{K}_H = \mathbf{K}_0 + \mathbf{H} \qquad [3]$$

Where, $\mathbf{K}_H$ and $\mathbf{K}_0$ are the wave vectors of the scattered and the incident beams; and $\mathbf{H}$ (H = 1/d *i.e.* the reciprocal of the lattice spacing of the diffracting planes under consideration) represents the periodicity of the crystal lattice.

The structure of the crystal is introduced in this formulation through the dielectric constant, κ, which is related to the electron density ρ. The ρ varies periodically with distance in the crystal. Therefore, κ varies periodically with the periodicity of the lattice. At any point **r**,

$$\kappa(\mathbf{r}) = 1 - r_e(\lambda^2/\pi)\,\rho(\mathbf{r}) \qquad [4]$$

Here, r_e is the classical radius of electron and λ is the wavelength of the X-ray. The electron density $\rho(\mathbf{r})$ can be expressed as a Fourier sum over the reciprocal lattice. Therefore, we can re-write Eqn [4] as:

$$\rho(\mathbf{r}) = (1/V)\Sigma_H F_H \exp(-2\pi i\mathbf{H}\times\mathbf{r}). \qquad [5]$$

Here, F_H is the structure factor. Putting this value of $\rho(\mathbf{r})$ in Eqn [4], we get

$$\kappa(\mathbf{r}) = 1 - \Gamma\,\Sigma_H F_H \exp(-2\pi i\mathbf{H}\times\mathbf{r}), \qquad [6]$$

Here Γ (= $r_e\lambda^2/\pi V$) is a constant which depends on the experimental conditions. It may be mentioned that the value of $\kappa(\mathbf{r})$ is very close to the unity, deviations being of a few parts per million.

A sum of plane waves is assumed as the solution of Maxwell's equations (1) and (2) in the form:

$$\mathbf{A} = \exp(2\pi i\nu l)\,\Sigma_H \mathbf{A}_H \exp(-2\pi i\mathbf{K}_H\times\mathbf{r}). \qquad [7]$$

Here, **A** stands for **E**, **D**, **B** or **H**. By combining Eqns [3] and [7] we get:

$$\mathbf{A} = [\Sigma_H \mathbf{A}_H \exp(-2\pi i\mathbf{H}\times\mathbf{r})]\exp(-2\pi i\mathbf{K}_0\times\mathbf{r})\exp(2\pi i\nu l). \qquad [8]$$

The formulation becomes considerable simple when only one reciprocal lattice point is active for diffraction, *i.e.* only this reciprocal lattice point is near the Ewald sphere. After some mathematical analysis, the fundamental set of equations for the wave field inside crystals can be written in the following form:

$$[k^2(1 - \Gamma F_0) - (\mathbf{K}_0.\mathbf{K}_0)]E_0 - k^2P\Gamma F_H E_H = 0, \qquad [9]$$

$$-k^2P\Gamma F_H E_0 + [k^2(1 - \Gamma F_0) - (\mathbf{K}_H.\mathbf{K}_H)]E_H = 0. \qquad [10]$$

In these equations, P represents the state of polarization of the X-ray beam and is one for σ polarization and its value is cos 2θ for π state. K is the value of wave vector of the X-ray beam in vacuum. The analysis of these equations yields expressions for the wave amplitudes as well as wave vectors. Geometrically, the relationship between different wave vectors is elegantly displayed by using the dispersion surface concept. In the following, we shall give expressions for intensities of diffracted X-ray beams for the reflection as well as the transmission geometries.

Reflection Geometry or Bragg Case

The expressions for X-ray intensities in the case of reflection and the transmission geometries are different from each other. The reflection geometry is known as Bragg case whereas the transmission configuration is labeled as Laue Case. As expected, in the case of Bragg geometry, the diffracted beam emerges from the same surface on which it is incident. For a centrosymetric crystal, one obtains the following expressions for reflection coefficient, defined as $|E_H^e/E_0 i|^2$

$$\frac{|E_H^e|^2}{|E_0^i|^2} = |b||\eta \pm (\eta^2 - 1)^{1/2}|^2 \qquad [11]$$

Here, η is a function of deviation from the Bragg angle and other basic parameters, which defines the experimental conditions including the diffracting planes under consideration. Since the real crystals have finite absorption coefficient for X-rays, the structure factor, the wave vectors, the dielectric constant and η are all complex quantities. For symmetrical Bragg geometry, value of b =–1 and the real and imaginary parts of h are given by:

$$\eta' = (-\Delta\theta \sin 2\theta + \Gamma F_0')/|P|\,\Gamma F_H', \qquad [12]$$

and $$\eta'' = -(F_H''/F_H')(\eta' - 1/|P|\varepsilon) \qquad [13]$$

By plotting reflection coefficient or reflectivity as a function of the deviation from the Bragg angle Δθ, one obtains theoretical diffraction curve for a desired set of diffracting planes and a given radiation in symmetrical Bragg geometry. It may be emphasized that this calculation is based on the plane wave dynamical theory.

Transmission Geometry or Laue Case

In the case of the transmission geometry, the following expressions are obtained for ratio of the intensities of the diffracted beam $I_{H\omega}^e$ and the forward diffracted beam $I_{0\omega}^e$ to that of the exploring beam I_0 are given by:

$$\frac{I_{0w}^e}{I_0} = \frac{1}{4}\left[1 \mp \frac{\eta'}{|1+(\eta')^2|^{1/2}}\right]^2 \exp\left[-\frac{\mu_0 t_0}{\gamma_0} 1 \mp \frac{|P|\varepsilon}{|1+(\eta')^2|^{1/2}}\right],$$

$$\frac{I_{0w}^e}{I_0} = \frac{1}{4}\frac{1}{1+(\eta')^2} \exp\left[-\frac{\mu_0 t_0}{\gamma_0}\left\{1 \mp \frac{|P|\varepsilon}{|1+(\eta')^2|^{1/2}}\right\}\right] \qquad [14]$$

Here, ω refers to a particular polarization state of either the α or the β branch of the dispersion surface, and t_0 is the thickness of the specimen crystal. The crystal has been assumed to be centrosymmetric and the upper negative sign is to be used when the active point is on the α branch on the dispersion surface whereas for active points of β branch, positive sign is to be used. In this case, the value of η is given by:

$$\eta' = (\Delta\theta \sin 2\theta)/ |P| \, \Gamma | \, F_H' | \quad [15]$$

For a very thin crystals $\mu_0 t_0 \cong 0$ and the total contribution of both the branches will be ½. These expressions are used to get theoretical diffraction curves for crystals. The absolute value of integrated intensity can be calculated by computing the area under the diffraction curve.

High Resolution X-Ray Diffraction: A Five Crystal X-Ray Diffractometer

The expressions derived in the foregoing Section were based on the assumption of an absolutely parallel and monochromatic exploring X-ray beam. However, such a beam is not realizable in actual practice. For real experiments, one is compelled to employ a beam, which has a finite angular divergence and wavelength spread. However, efforts are made to minimize these parameters while keeping under consideration that experiments have to be completed in reasonable time. In many experiments, particularly, those dealing with the investigation of low dimensional materials and devices, the dimensions of the exploring beam are also required to be small.

To achieve a nearly parallel and monochromatic X-ray beam, one employs diffraction from high quality crystals, which is limited to a narrow range of angles and there is very high selectivity in wavelength of X-rays. A variety of multi-crystals X-ray diffractometers have been developed to achieve these goals (Bowen and Tanner, 1998; Lal, 1998; Zachariasen, 1945; Pinsker, 1978; Authier, 2001). In author's laboratory, a series of multi-crystal X-ray diffractometers have been developed over the years for different high-resolution X-ray diffraction experiments (Lal, 1998).

These include a Five-Crystal X-ray Diffractometer with state-of-the art-level resolution. A brief description of this system is given in the following.

Figure 7.1 shows a schematic diagram of the Five-Crystal X-ray Diffractometer designed, developed and fabricated at author's laboratory (Lal, 1993; Lal, 1998). X-ray beam emerging from a fine-focus or a micro-focus X-ray source is mechanically collimated with the help of a long collimating tube (30-50 cm), which is fitted with a pair of fine slits, the slit opening can be adjusted to have desired width.

The collimator is generally oriented for a small angle of foreshortening (~ 3°). The combination of an X-ray source of small size, a long collimator and fine slits gives an X-ray beam whose divergence in the plane of diffraction is typically about 80 arc sec. This beam is diffracted from a set of three plane silicon monochromator-collimators oriented in (+, -, +) geometry. The first two crystals have their surfaces along (111) set of planes and are of Bonse-Hart type (Bonse and Hart, 1965) set in (+,-) geometry. The X-ray beam diffracted from these crystals consists of well-resolved

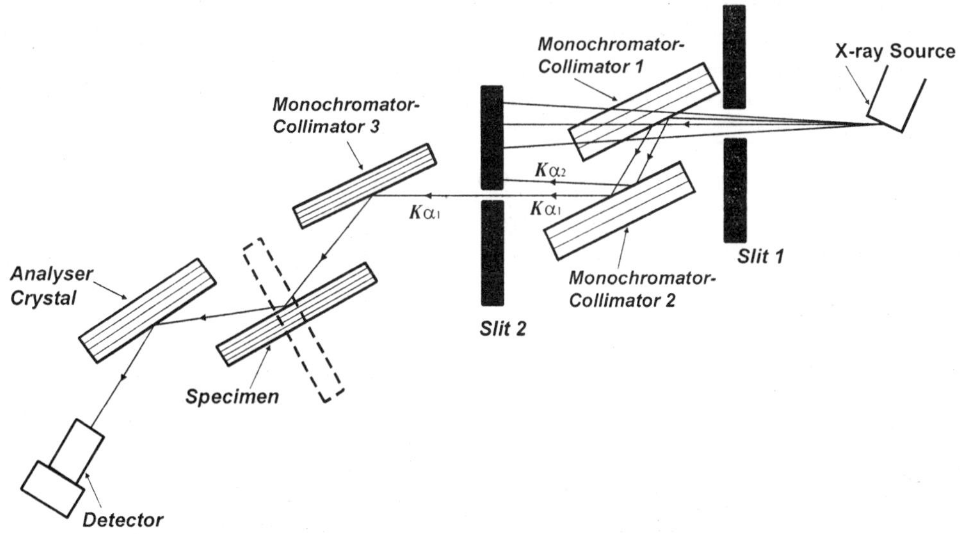

Figure 7.1: A Schematic Line Diagram of the Five Crystal X-ray Diffractometer Designed, Developed and Fabricated at Author's Laboratory

$K\alpha_1$ and $K\alpha_2$ beams when a fine focus source is used and either $K\alpha_1$ and $K\alpha_2$ component is obtained, if a source with small size [~0.1mm or less] is employed. The $K\alpha_1$ beam is isolated with the help of another slit and is diffracted from the third crystal, which is oriented for diffraction from planes like (800) to achieve a high level of dispersion. Diffraction from third crystal leads to a high reduction in the wavelength and direction spreads of the X-ray beam. Both the spreads together add up to much less than 0.3 arc sec. The specimen is generally placed at the fourth crystal position and the fifth crystal serves as an analyzer.

The third, the fourth and the fifth crystal stages have provisions for tilting, translation and θ (glancing angle) rotations. A vertical circle goniometer allows full circle azimuthal rotations of specimens. Scintillation counters are used as X-ray detectors. The output of these detectors is fed to counting systems, which can be connected to a PC. Also, the glancing angle θ can be adjusted by stepping motors, which are controlled by the PC. The diffractometry experiments can be performed either manually or automatically with the help of the PC. The diffractometer can be used for performing the following main experiments: (*i*) diffractometry; (*ii*) topography, (*iii*) diffuse X-ray scattering measurements; (*iv*) curvature measurements; (*v*) determination of crystallographic orientation; (*vi*) measurement of lattice parameters of single crystals; (*vii*) determination of lattice and orientational mismatches between epitaxial films and substrate single crystals; and (*viii*) measurement of absorption of crystals for X-rays.

Figure 7.2(a) shows a typical diffraction curve of a nearly perfect (111) silicon single crystal recorded with (111) diffraction planes and $MoK\alpha_1$ radiation by using (+, -, +, -, +) geometry of the Five-Crystal X-ray Diffractometer (Lal, 1993; Lal, 1998).

This curve is an as-recorded experimental data without any corrections. The theoretical diffraction curve obtained by using plane wave dynamical theory for the same experimental conditions is shown in Figure 7.2(b) (Lal, 1993). A comparison of the two curves shows that the experimental curve is very close to the theoretical curve.

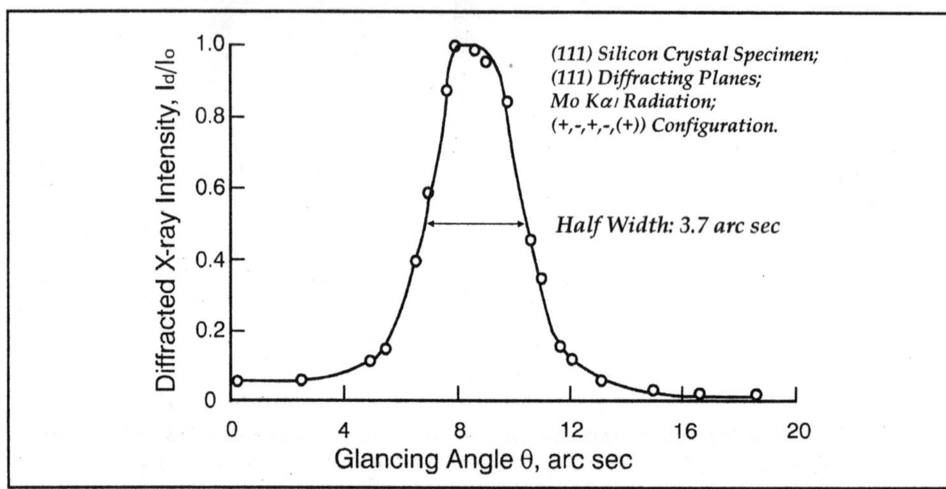

Figure 7.2(a): A High Resolution X-ray Diffraction Curve of a (111) Silicon Single Crystal Recorded on the Five Crystal X-ray Diffractometer in Symmetrical Bragg Geometry and (+,–, +,–, +) Configuration of the Diffractometer. Mo Kα_1 exploring beam was employed. The points represent the as measured data without any correction being applied.

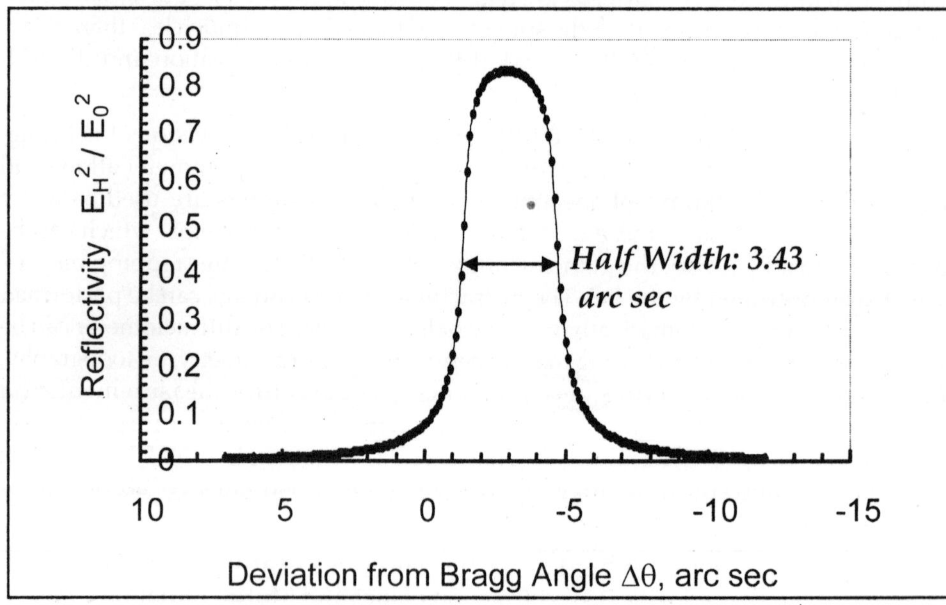

Figure 7.2(b): A Theoretical Diffraction Curve Calculated on the Basis of Plane Wave Dynamical Theory for an Ideally Perfect Silicon Crystal for Experimental Conditions of (a)

The half width of the experimental curve is larger than that of the theoretical curve by only ~ 0.3 arc sec. This deviation comprises of the instrumental broadening, the residual defects in the specimen and the wavelength and direction spreads of the exploring X-ray beam. If we record diffraction curve of a higher order reflection like 400 without changing the monochromators, the dispersion broadening still remains negligible. For example, the typical half width of an experimental diffraction curve was only ~ 1.7 arc sec, as compared to the theoretical expected value of ~ 1.4 arc sec. This indicates the high resolution available with this system. The author and his colleagues have employed this and the other diffractometers for a variety of fundamental and applied research. Some important results are described in the following.

Characterization of Defects in Nearly Perfect Crystals

Following are important crystal defects or lattice imperfections about which information is essential in fundamental understanding of properties of crystals, for repeated preparation as well as for their conversion into devices: boundaries, dislocations, point defect aggregates and isolated point defects. High-resolution X-ray diffraction techniques can be used for direct observation of most of the important defects like boundaries and dislocations in a non-destructive manner. Even point defects and their clusters can be detected and studied by employing high-resolution diffuse X-ray scattering measurements. Some recent interesting results on defect studies are described in the following.

Characterization of Very Low Angle Boundaries

Generally, a low angle boundary is expected to separate two regions of a single crystal, which have an angular tilt of ~ 1 arc min or more. Recently, we have observed low angle boundaries, which have much smaller tilt angles. These have been observed and characterized in single crystals of bismuth germanate (BGO) (Lal, 1998; Zachariasen, 1945; Laue, 1960), cadmium telluride and mercury cadmium telluride. These boundaries have been termed as very low angle boundaries.

Figure 7.3 shows a typical diffraction curve of a region near the periphery of a BGO crystal, which contains a *very* low angle boundary (Murthy *et al.*, 1999; Choubey, 2000; Choubey *et al.*, 2002).

This curve consists of two well-resolved peaks, which are separated by an angle of 33 arc sec.

Each one of the curves is quite sharp with half width of ~8 arc sec, which is not far from the theoretical value of the half width of 3.8 arc sec. High-resolution X-ray diffraction traverse topographs were recorded after orienting the specimen at peaks 1 and 2 in Figure 7.3 and these are respectively shown in Figures 7.4 (a) and (b) (Murthy *et al.*, 1999; Choubey, 2000). It is clearly observed that complementary areas of the crystal are diffracting in Figures 7.4 (a) and (b). We have investigated *very* low angle boundaries in BGO crystals of varying degrees of perfection (Choubey, 2000; Choubey *et al.*, 2002). It has been observed that the perfection of a crystal containing several low angle boundaries improves significantly on thermal annealing and the boundaries can be moved outside the crystal. However, often, it was observed that a few *very* low

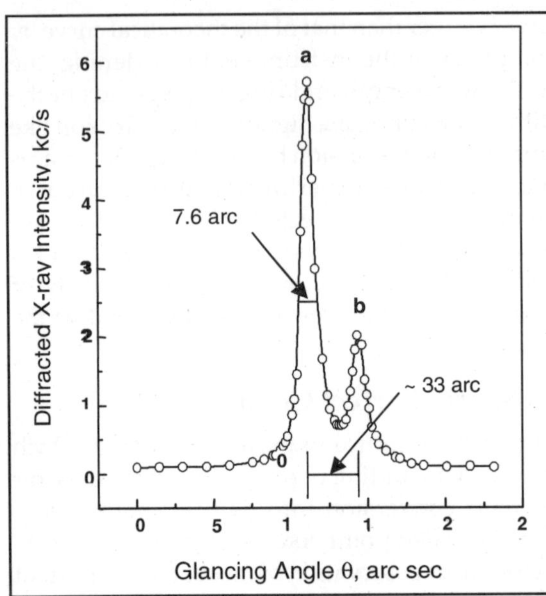

Figure 7.3: A High Resolution X-ray Diffraction Curve Recorded with (400) Planes of a (100) Bismuth Germanate (BGO) Crystal in Symmetrical Bragg Geometry and Mo Kα_1 Radiation. The area of the crystal under investigation was close to its periphery, where presence of a low angle boundary was indicated by topographic examination of the crystal.

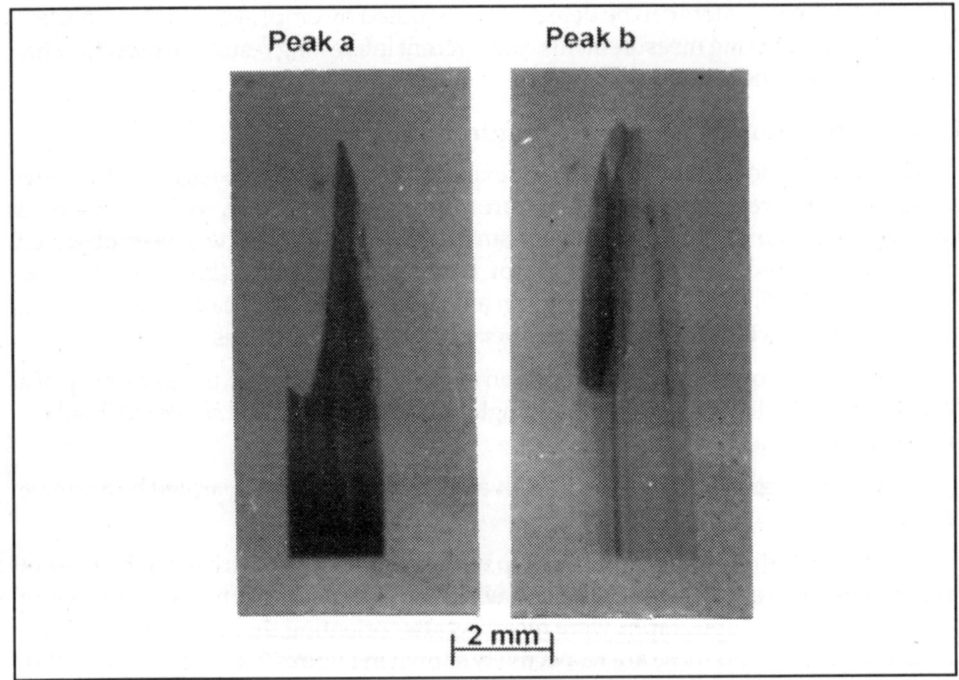

Figure 7.4

(a) A HRXRD traverse topograph of the specimen of Figure 7.3. It was recorded by orienting the specimen on the more intense peak in Figure 7.3. Other experimental conditions were the same as in Figure 7.3.

(b) A HRXRD traverse topograph of the specimen of Figures 7.3 and 7.4 (a) recorded by orienting the specimen at the peak position of the weaker diffraction maximum in Figure 7.3.

angle boundaries could not be annealed out even when the specimens were heated at temperatures that were very close to their melting point. An examination by Secondary Ion Mass Spectrometry (SIMS) revealed that the region of the immovable *very* low angle boundary had high concentration of impurities like silicon (Choubey *et al.*, 2002). Apparently, the impurities decorate the defect and do not allow it to move out on annealing.

Direct Observation and Characterization of Dislocations in Nearly Perfect Crystals

X-ray diffraction topography is widely employed as a non-destructive direct method for imaging dislocations in nearly perfect crystals (Lang, 1958; Bowen and Tanner, 1998; Kato, 1992; Lal, 1993). When multi-crystals X-ray diffractometers are employed, the sensitivity to the strain field around dislocations is enhanced significantly. This makes it possible to image a large region of strain field around the dislocation. Figure 7.5 shows a high-resolution traverse topograph of a GaAs crystal grown by horizontal Bridgman method (Lal *et al.*, 1990). In this topograph, only three dislocations are observed. The size of each image is about a millimeter, showing the high sensitivity of the experimental technique. It may be mentioned that the half width of the diffraction curve of this specimen was ~10 arc sec and the specimen was required to be maintained at the peak of the diffraction curve while it was traversed across the exploring X-ray beam. Similar results have been observed with a variety of crystals like those of lithium niobate, sapphire, diamond, garnets and others. In the case of silicon, it is very rare to find a dislocation.

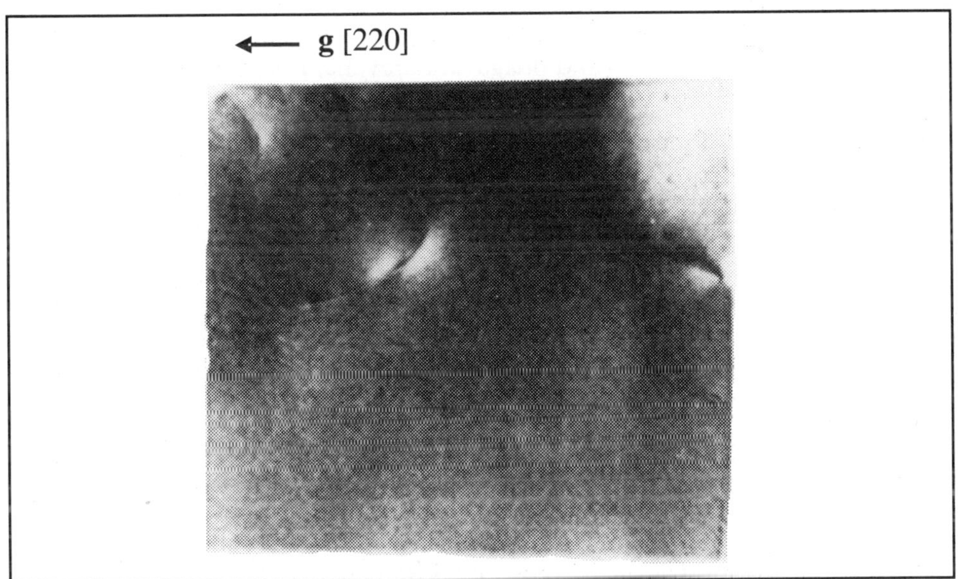

Figure 7.5: A HRXRD Traverse Topograph of a Nearly Perfect Gallium Arsenide Crystal Recorded in Symmetrical Laue Geometry with Mo Kα_1 Radiation. Only three dislocations are observed in the entire volume of the wafer.

Characterization of Point Defects and their Clusters

Single crystals, which are free from boundaries and dislocations, cannot be considered as ideally perfect as these contain varying degrees of point defects and their clusters. The author and his co-workers have employed high-resolution diffuse X-ray scattering (DXS) measurements to study point defects and their clusters in nearly perfect crystals (Choubey *et al.*, 2002; Lal *et al.*, 1979; Lal and Singh, 1980; Lal and Singh, 1981; Singh, 1980; Lal *et al.*, 1991; Lal *et al.*, 1996; Bhagavannarayana, 1994; Ramanan *et al.*, 1995; Lal, 1989; Ramanan *et al.*, 1998; Lal *et al.*, 2000; Ramanan, 1998; Bhagavannarayana *et al.*, 2005). In the first phase, a systematic study was carried out to understand the origin of diffuse X-ray scattering from crystals (Lal *et al.*, 1979; Lal and Singh, 1981; Singh, 1980). Till early seventies it was believed that elastic thermal waves or phonons are primarily responsible for diffuse X-ray scattering around diffraction maxima (Wooster, 1962). The technique of making DXS measurements was improved substantially by employing highly collimated and monochromated $K\alpha_1$ X-ray beams for exploring reciprocal space very close to the reciprocal lattice points and at high resolution (Lal *et al.*, 1979). Diffuse X-ray scattering measurements were made on a variety of crystals having different Debye temperatures and around different reciprocal lattice points. Experiments were also performed at elevated temperatures. The results of these studies showed that at and near room temperature, the diffuse X-ray scattering is primarily due to defects and not due to thermal vibrations of the crystal lattice.

DXS measurements have been employed to investigate point defects and their clusters in a variety of as-grown crystals. In particular, detailed studies have been carried out on device quality silicon single crystals grown by Czchoralski (CZ) as well as Float-zone (FZ) methods (Bhagavannarayana, 1994; Ramanan *et al.*, 1995; Lal *et al.*, 2000; Ramanan, 1998; Lal and Bhagavannarayana, 1989).

The purity of these crystals has also been varied. The effect of thermal annealing under controlled conditions has also been studied (Lal *et al.*, 2000; Ramanan, 1998). Also, investigations of process-induced defects like those due to ion-implantation in silicon have yielded very interesting results concerning segregation of implanted impurity (Lal *et al.*, 1991; Lal *et al.*, 1996). Studies on BGO crystals have yielded interesting results (Ramanan, 1998; Bhagavannarayana, 2005). This technique has been enlarged in recent years and is sometimes referred to as reciprocal lattice mapping. Here, we shall describe some interesting results on point defects and their clusters in silicon crystals and the changes in the same due to annealing under oxygen at different elevated temperatures.

Investigations of point defects and their clusters in silicon crystals of varying degrees of purities and grown by Czchoralski (CZ) and Float-zone (FZ) methods have shown that vacancy type defects give dominant contributions to DXS in the case of crystals grown by the FZ method (Lal and Bhagavannarayana, 1989; Bhagavannarayana, 1994; Ramanan *et al.*, 1995; Ramanan, 1998). In the CZ crystals, the dominant defects are of interstitial type. In both the cases, isolated as well as the agglomerated point defects have been observed. The experimental data could be analyzed to deduce important parameters of defects like cluster radius by using a phenomenological model proposed by Dederichs (Dederichs, 1971) for crystals containing low density of dislocations loops formed by loosely clustered point defects. It was found that the radii of the clusters were slightly different in the FZ and CZ grown crystals. Typical values are ~0.6 μm for the vacancy clusters and 0.8 μm for the interstitial clusters (Ramanan *et al.*, 1995; Ramanan, 1998). It has been observed that the DXS distribution is very sensitive to the processing of crystals and their thermal history. In particular, DXS intensity from regions very close to the reciprocal lattice point is a very sensitive function of the real structure of crystals.

As mentioned above, there is a significant difference between the point defects and their clusters, which produce diffuse X-ray scattering around reciprocal lattice points of the FZ and the CZ grown silicon crystals. It was also observed that in general, the overall degree of perfection of the CZ crystals was slightly better than that of the FZ crystals (Lal *et al.*, 1991; Bhagavannarayana, 1994; Ramanan *et al.*, 1995; Ramanan, 1998). The half width of the diffraction curves of the CZ crystals was smaller by a few arc sec or so in comparison to that for the FZ crystals. Also, the absolute DXS intensity was higher for FZ crystals at all values of K* in comparison to that of the CZ crystals. K* is a vector, which joins the elemental volume of reciprocal space under investigation to the nearest reciprocal lattice point. DXS intensity is a good indication of the overall degree of perfection of the crystals. This result essentially means that relatively oxygen free FZ silicon crystals have lower degree of perfection in comparison to the CZ grown silicon crystals, which have appreciably higher concentration of oxygen. In the case of natural diamond crystals also, it is known that small concentration of nitrogen has a strong influence on the degree of perfection of these crystals

We have performed experiments by deliberately increasing the concentration of oxygen in high purity silicon crystals grown by the float zone method. The concentration of oxygen in the silicon specimens was monitored by measuring infrared absorption at 1105 cm^{-1} (ASTM, 1980; Bergholtz *et al.*, 1989). When the concentration of interstitial oxygen increased, there was a shift in the absorption peak from 1105 cm^{-1} to 1070 cm^{-1} suggesting precipitation of oxygen. A band at 470 cm^{-1} also appears when SiO_2 precipitates are present in appreciable concentration. The oxygen concentration increases from 1.3 ´ 10^{17} cm^{-3} in the as-grown crystals to 3.6×10^{17} cm^{-3} after annealing at 1073 K. Good quality CZ grown crystals generally have this level of oxygen. After annealing at elevated temperatures, the oxygen concentration increased substantially. In the specimen annealed at 1373 K, the concentration was 1.7×10^{18} cm^{-3}.

High-resolution X-ray diffractometry, absolute integrated intensity ρ measurements and DXS measurement experiments were performed after each annealing step. The half width of the diffraction curves of the specimen were typically ~ 11 arc sec before any heat treatment as shown in Figure 7.6. No significant change in the real structure or degree of perfection was observed up to annealing temperatures of 873 K. Between 873 K and 1073 K, remarkable changes were observed in diffraction curves as well as in the DXS intensity distribution. Figure 7.6 shows three diffraction curves of this crystal. These curves were recorded before annealing, after annealing at 1073 K and after annealing at 1373 K. It is found that there is a remarkable decrease

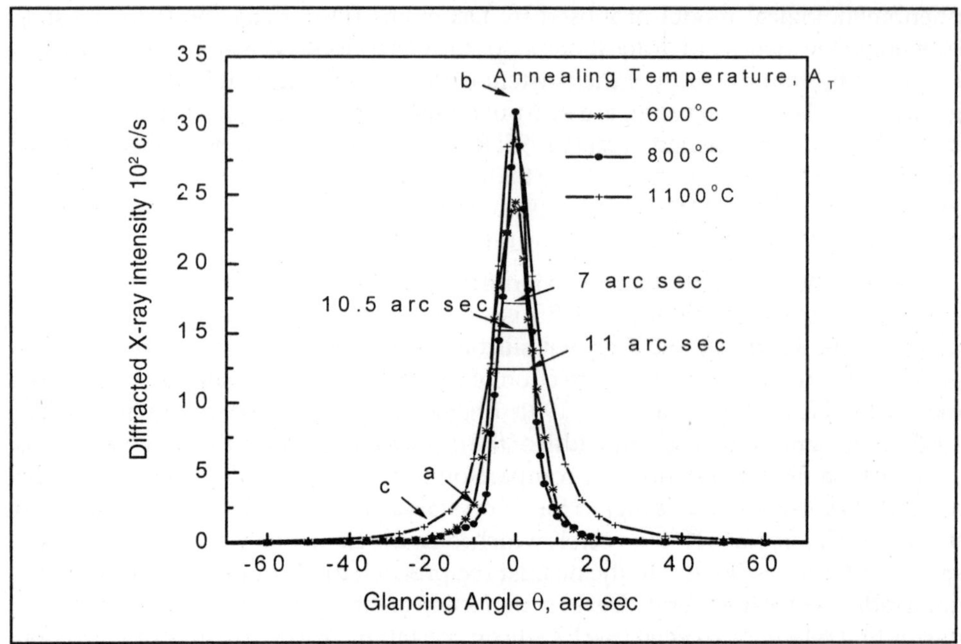

Figure 7.6: Typical Diffraction Curves of a Nearly Perfect Single Crystal of Silicon. Mo Kα_1 radiation, symmetrical Bragg geometry and (+,–, +,–) configuration of the diffractometer were employed. (a), (b) and (c) were recorded after annealing at 873 K, 1073 K and 1373 K, respectively.

in the half width of the diffraction curve to 7 arc sec after annealing at 1073 K. However, it again increases to 10.5 arc sec after annealing at 1373 K. The oxygen concentration at 1073 K was 3.6×10^{17} atoms cm^{-3}, which is nearly the same as that observed in good quality CZ grown crystals. After annealing at 1373 K the oxygen content increases substantially.

The results of absolute integrated intensity ρ measurements also support conclusions of diffractometry experiments. The value of ρ decreases from 3.5×10^{-5} rad (before treatment) to 2.4×10^{-5} rad after annealing at 1073 K. However, it increases to 4.3×10^{-5} rad after annealing at 1373 K. This experiment indicates a direct correlation between the concentration of oxygen in these crystals and their degree of perfection. The degree of perfection is highest at oxygen levels of ~ 4×10^{17} cm^{-1}.

Figure 7.7 shows a typical set of four DXS intensity versus K* plots recorded with a specimen at the following four stages: (*i*) before annealing; (*ii*) after annealing at 873 K; (*iii*) after annealing at 1073 K; and (*iv*) after annealing at 1373 K. In Figure 7.7 we show only those plots in which K* was along the reciprocal lattice vector R* for (111) planes. Measurements had been made with the K* along [1 1 1], ['1'1'1], ['1 1 0] and [1 '1 0] directions. The defects, particularly the clusters could be modelled as platelets lying on (111) planes. Therefore, for the sake of clarity, only data with K* along ['1'1'1] {curve (a), vacancy type defects} and [1 1 1] {curves (b), (c) and (d),

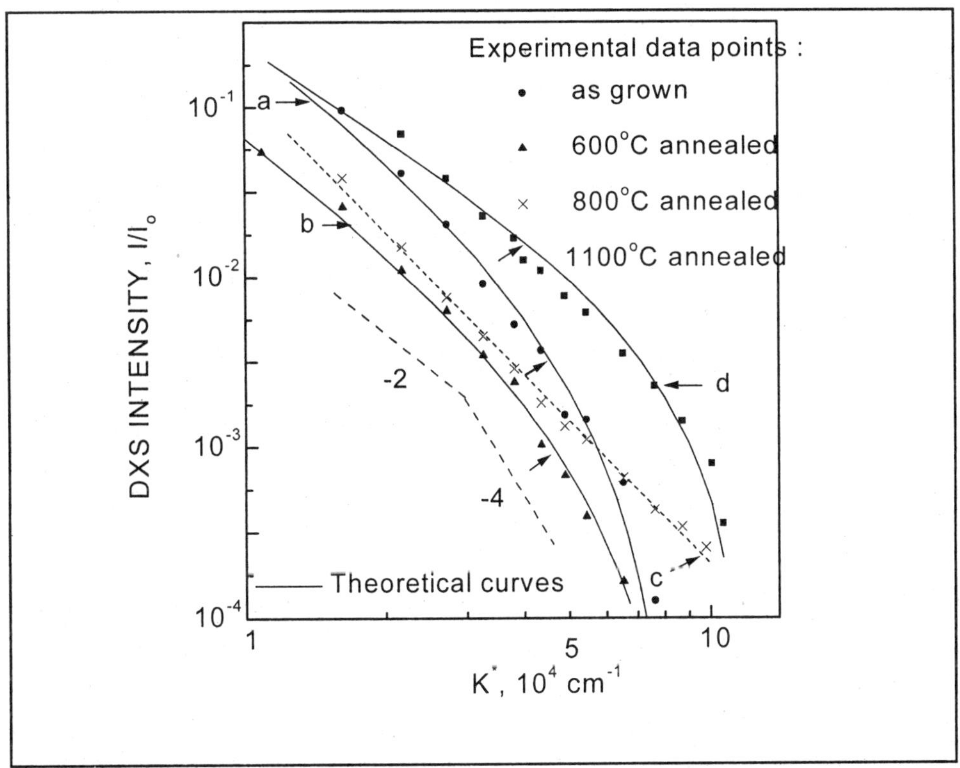

Figure 7.7: A Set of Typical Log [DXS intensity I/I_0] vs. log K* Plots for a Silicon Crystal Grown by the FZ Method. (a), (b), (c) and (d) are respectively for as grown state and after annealing at 873 K, 1073 K and 1373 K. Other experimental conditions were the same as in Figure 7.6.

interstitial type defects} is shown in this figure. The points in these plots represent the experimental data. The data was analyzed with the help of a phenomenological model developed Dederichs (Dederichs, 1973) and full lines were obtained by this modeling exercise. It can be seen that the fit between the experimental data points and the theoretical curves is quite good. It may be mentioned that no significant change in diffuse X-ray scattering distribution was observed up to annealing temperature of 723 K. In the as grown state, the DXS data showed the dominance of isolated vacancies and their clusters. The cluster radius was ~ 0.2 µm. A comparison of curves (a) and (b) clearly shows that there is an overall decrease in DXS intensities at all values of K*. It may be mentioned that while computing theoretical intensity a scaling factor C has to be used. C depends upon the experimental conditions such as wavelength, direction and spatial spreads of the exploring beam, the width of the slit in front of the detector and the density of defects in the crystal. The value of C was smaller for curve (b) in comparison to that for curve (a) by ~ 20 per cent.

Since the experimental conditions were identical, the decrease in the magnitude of the scaling factor is essentially due to corresponding decrease in the defect density. A small decrease in the size of the clusters has also taken place (from 0.24 µm to 0.22

µm). Most remarkable changes take place after annealing the crystals at 1073 K {curve (c)}. The sign of anisotropy of the DXS distribution changes from negative to positive, indicating that the main source of diffuse X-ray scattering is now defects of interstitial type and not vacancies. The curve (c) is essentially a single straight line with slope of –2 throughout the observed K* region. This shows the vacancy clusters present in the as grown crystals have dissociated and the dominant contribution is from the isolated interstitials, which could be oxygen impurity atoms. In curve (d), we again see two well defined regions in the plot with slope of –2 in the smaller value range of K* and slope of –4 in the higher K* value region. This indicates clustering of interstitials, whose size has been deduced as ~ 0.8 µm.

The value of the scaling factor C has also increased by ~ 30 per cent in comparison to that obtained with this very crystal before annealing.

The combination of the results obtained by diffractometry, measurement of absolute integrated intensity and analysis of DXS distribution clearly indicates that the perfection of the specimen is dependent on an optimum concentration of oxygen in silicon crystals. An oxygen concentration of ~ 4×10^{17} cm^{-3} without significant segregation is optimum for high degree of perfection of silicon crystals.

Structural Characterization of Thin Epitaxial Films and Determinational of Biaxial Stress in Thin Film-Single Crystal Substrate Systems

Thin films of a wide variety of materials are extensively employed in almost all devices. It is a challenge to unravel the real structure of these films when one or more dimensions are restricted to very small values, sometimes in nano-meter range. X-ray diffractometry, particularly in the grazing incidence mode has been quite effectively used (Stepanov *et al.*, 1996). Even normal high resolution X-ray diffractometry gives very valuable information about epitaxial films regarding crystallinity, structural quality, lattice mismatch between the films and the substrates and composition of films as deduced from lattice mismatch. In addition, biaxial stress in thin film – single crystal substrate systems strongly influence their properties and affect their long-term mechanical stability. Biaxial stress is produced due to dissimilar lattice parameters of the films and the substrates and due to difference in thermal coefficients of expansion of the thin film and the substrate material. The later is an important reason as the films are generally deposited at elevated temperatures and later cooled to room temperature. The value of stress has to be controlled to ensure a control on physical properties as well as for long-term stability of the thin film-substrate system. If the stress is very high, the films simply peel off. Therefore, it is important to be able measure the stress accurately, understand the key parameters that control stress level and to optimize experimental conditions to minimize stress. A wide variety of thin films had been investigated in author's laboratory. These include: MBE grown structures of InAlAs, InGaAs, SiGe, CdTe, GaSb and films grown by other techniques like those of silicon oxide, silicon nitride, polysilicon, silicon oxynitride and molybdenum silicide on silicon and multi-level metallizations on gallium arsenide. Some typical results of structural characterization and biaxial stress determination are described in the following.

Structural Characterization of MBE Grown Thin Epitaxial Films

Strained layers of SiGe on silicon have been investigated extensively because of their exciting properties like optical absorption and their possible use as radiation detectors (Metzger, 1995; Kato *et al.*, 1995). Molecular Beam Epitaxy (MBE) under ultra high vacuum conditions has been widely used to prepare SiGe epitaxial films. The film thicknesses are quite small, of the order of 30-50 nm. These are generally grown on (001) silicon substrates. High resolution X-ray diffractometry could be successfully employed to investigate the structural quality of SiGe films (Lal, 1995; Mukhopadhyay *et al.*, 1997). Figure 7.8 shows a typical diffraction curve of a 40 nm thick SiGe film. It is observed that a well-defined peak is obtained by employing normal incidence X-ray diffraction even though the film thickness is quite small. From the experimental value of angular separation between the peaks of film and the substrate, the value of $\Delta d/d$ was determined as 0.0103. The concentration of germanium could be determined from this value as 17.6 per cent . Therefore, the composition of the film was determined as $Si_{82.4}G_{17.6}$. It may be mentioned that lattice mismatch is not the only mismatch between the film and the substrate. In general, there is an additional mismatch due to a finite angle between the lattice planes of the substrate and those of the epitaxial film. By employing the fifth crystal (analyzer) of the diffractometer, it can be determined quantitatively. When the analyzer crystal is employed, the separation between the film and substrate peaks is only due to the lattice mismatch. However, without the analyzer crystal, the observed mismatch is both due to the usual lattice mismatch as well as due to the orientational mismatch. By combining two sets of measurement, with and without the analyzer crystal, it is possible to determine the value of both the mismatches accurately.

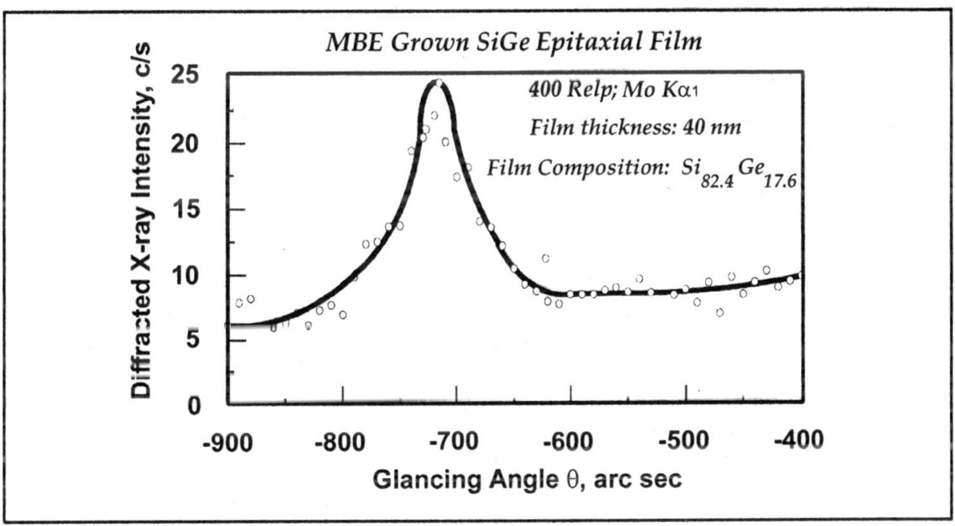

Figure 7.8: A Typical High Resolution X-Ray Diffraction Curve of a 40 nm Thick Silicon-Germanium Film Grown by Molecular Beam Epitaxy on a (100) Silicon Substrate

Determination of Biaxial Stress in Thin Film-Single Crystal Substrate Systems

High-resolution X-ray diffraction techniques have been widely used to determine biaxial stress in thin film – substrate single crystal systems (Lang, 1958; Mukhopadhyay *et al.*, 1997; Rozgonyi and Ciesielka, 1973). In author's laboratory, a variety of systems have been investigated.

These include: (i) SiO_2 on Si; (ii) SiO_2 + Si_3N_4 on Si [47]; (iii)

Si oxynitride on Si (Halder *et al.*, 1994); (iv) LTO + polysilicon on Si (Goswami *et al.*, 1996); (v) $MoSi_2$ on Si (Lal *et al.*, 1996); and (vi) multi-level metallizations on GaAs (Lal *et al.*, 1990; Goswami *et al.*, 1994). This work has led to interesting new results.

It is well known that when the film-substrate system is stressed it leads to the bending of the system. The biaxial stress σ is related to the radius of curvature R, the thicknesses of the film t_f and the substrate t_s, and the elastic modulus E and the Poisson ratio ν of the substrate by the following expression (Rozgonyi and Ciesielka, 1973):

$$\sigma = [E/6(1-\nu)]\,(t_s^2/t_f)\,(1/R) \qquad [16]$$

If one can measure the value of R the value σ can be determined from Eqn. [16]. We had employed high-resolution X-ray diffraction techniques to measure the curvature of the substrate crystal. Since the diffraction curves of the semiconductor crystal substrates are quite sharp, the radii of curvature of several tens of kilometers can be measured accurately.

Systematic study of biaxial stress in a number of systems had revealed that the relation given in Eqn. [16] is not adequate to express the relationship between the stress and other parameters. This relation assumes that the starting wafer or the substrate is absolutely flat.

However, no real substrate is flat. These have varying degrees of bending and can have in-built tensile or compressive stresses. In general, the situation is like that shown in Figures 7.9(a) and (b). In Figure 7.9(a), the starting wafer is concave in shape when viewed from the polished side on which the film is deposited. If the film produces a tensile stress, the wafer will remain concave but its radius of curvature will decrease or it's bending will increase. If the stress is compressive and small in value, the wafer will still remain concave but its radius of curvature will increase. If the compressive stress is quite high, it will change its shape and it will become convex as shown in the figure. We can see that if the initial bending of the wafer is not taken into consideration, the value of stress deduced from the radius of curvature will be in error. Also, the nature of the stress may be wrongly inferred. Similarly, if the starting wafer is convex in shape, the possible shapes of the wafer due to compressive and tensile stresses are shown in Figure 7.9 (b). To take into account the initial bending of the wafer, we have employed the following relation (Lal *et al.*, 1996):

$$\sigma = [E/6\,(1-\nu)]\,(t_s^2/t_f)\,\{(1/R)-(1/R_o)\} \qquad [17]$$

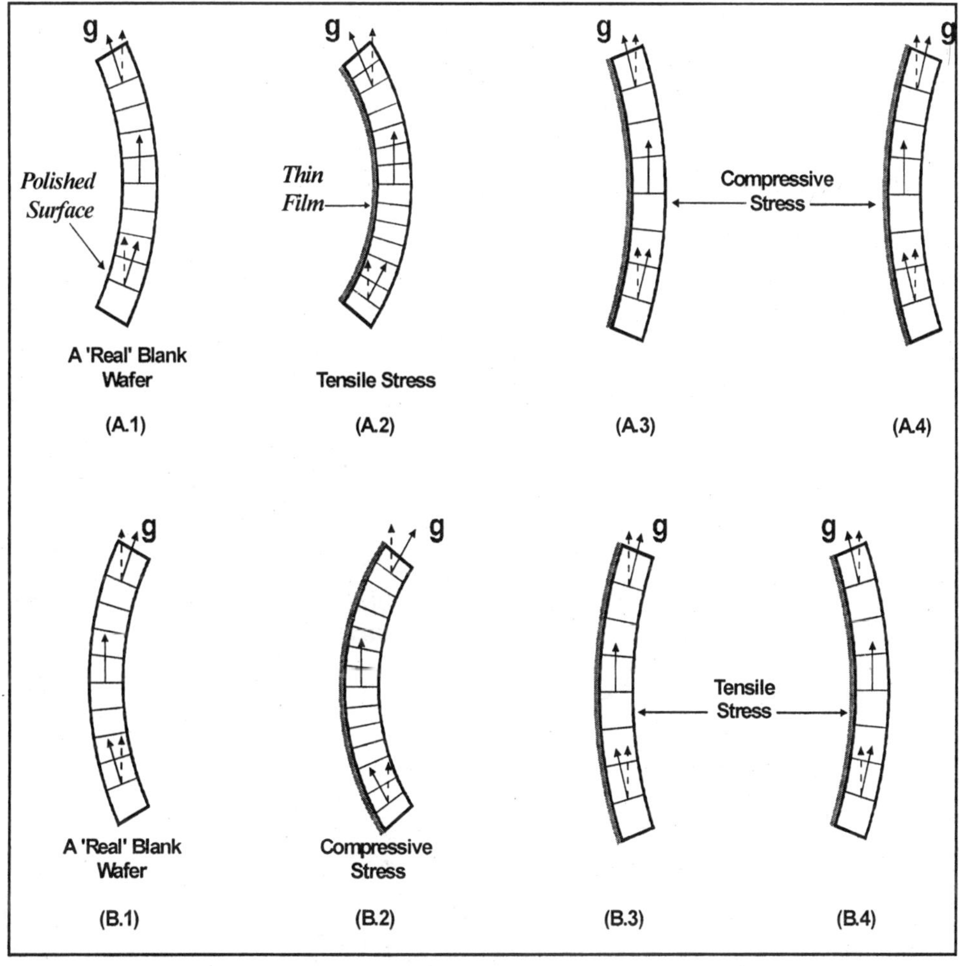

Figure 7.9

(a) A schematic line diagram showing bending produced by deposition of a thin film on a concave shaped substrate.

(b) The bending of the substrate due to stress produced by a thin film when the wafer was convex shaped to begin with.

Here, R_o is the radius of curvature of the blank wafer. By recording diffraction curves at different points of the specimen and by using X-ray diffraction topography, one can directly observe the effect of stress on crystalline perfection and distribution of stress in the substrate crystal (Lal *et al.*, 1990; Halder *et al.*, 1994a; Halder *et al.*, 1994b; Goswami *et al.*, 1996; Lal *et al.*, 1996).

We describe here an interesting result obtained with $MoSi_2$ films on Si substrates (Lal *et al.*, 1996). Molybdenum silicide films (100 nm thick) with tetrahedral structure were deposited by co-sputtering followed by rapid thermal annealing. The substrates were (100) silicon wafers with thickness of 300 μm. In this study, not only the value of stress was determined but also the effect of initial bending of the substrate on the final stress value was studied. Also, the effect of thermal annealing on the stress was studied to optimize annealing conditions for lowering the stress. Figure 10 shows three curvature plots. The specimen substrate crystal was oriented for diffraction in symmetrical Laue geometry. It was traversed across the X-ray beam in small linear steps. If the specimen is absolutely flat, it will remain on the peak of the diffraction maximum at all linear positions. However, all real crystals are bent and as these are moved across the X-ray beam, these need to be reoriented to attain the diffraction peak position. The rotation provided after each translation is measured experimentally and is plotted as a function of the linear position, as in Figure 7.10. These plots, termed as curvature plots have a positive slope if the substrate is convex in shape, when viewed from the polished side. The slope is negative if the substrate shape is concave. The value of radius of curvature R is determined from the slope of the

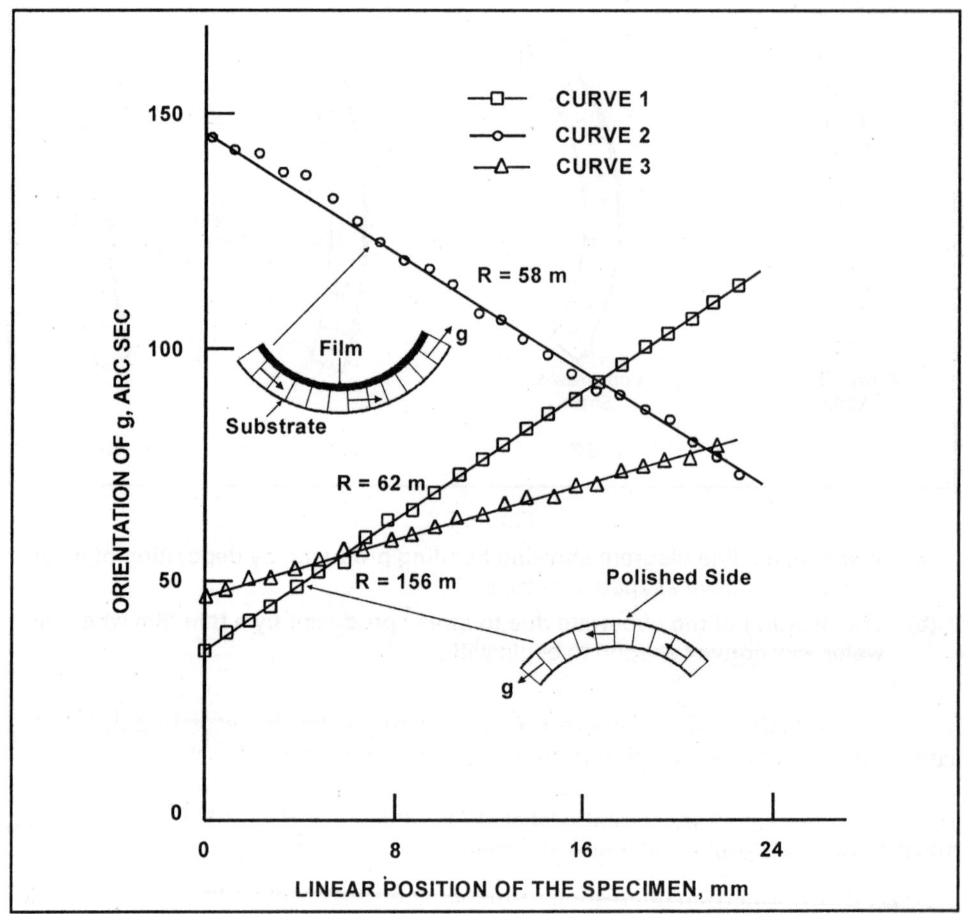

Figure 7.10: Typical Curvature Plots for a Silicon Substrate on which a 100 nm Thick $MoSi_2$ Film was Deposited by Co-sputtering. The results are for the blank specimen, just after deposition of the film and after annealing the crystal with film.

curvature plot. It can be seen from Figure 7.10 that the blank wafer was convex when viewed from the polished surface side. It had rather low value of radius of curvature (62 m) showing a high level of bending. After the deposition of the film, the shape of the substrate changes and it became concave in shape with a radius of curvature of ~ 58 m. After rapid thermal annealing at 1273 K, the shape the substrate changed again and it regained its original convex contour. However, its radius of curvature increased to 156 m. The value of stress was found to be -9.04×10^8 Nm^{-2} (tensile) in the as deposited system. After annealing, the stress reduced substantially to -2.63×10^8 Nm^{-2}.

We had investigated the effect of initial bending of wafers on the final value of stress. In this investigation, highly bent wafers with very small value of radius of curvature of 33m, as well as wafers with low level of bending wafers (radius of curvature ~ 250 m) were selected. Wafers with radii of curvature similar to that whose results are reported above, were also investigated. It was found that the initial bending had a very strong influence on the final value of biaxial stress.

The results of this study are given in Table 7.1.

Table 7.1: Some Typical Values of Radii of Curvature and Biaxial Stress in Three Specimens of $MoSi_2$ Films on Silicon Substrates

Sample No.	*Radii of Curvature m*			*Biaxial Stress 10^8, Nm^{-2}*	
	Blank Wafer	*As-deposited*	*Annealed*	*Blank Wafer*	*Annealed*
3	33	–49	114	–13.73	–5.83
2	62	–58	156	–9.04	–2.71
4	250	–121	275	–3.31	–0.10

It is seen from Table 7.1 that the value of stress may range from -13.73×10^8 Nm^{-2} (for the as deposited wafer with R_o = 33 m) to -0.10×10^8 Nm^{-2} (for the wafer with R_o = 250 m after optimum annealing). Hence, we see the stress can be reduced by more than 100 times by proper selection of the starting wafer and subjecting the wafers to annealing at optimized temperatures. In the case of GaAs crystals, we have seen that even surface preparation affects the final value of stress. In this case also the stress could be reduced by more than a factor of hundred ((Lal *et al.*, 1990; Goswami *et al.*, 1994).

Acknowledgement

The author had great privilege and pleasure in collaborating with a number of colleagues and students in his laboratory and with a number of distinguished scientists from India and abroad.

He would particularly like to acknowledge collaboration with Drs. Niranjana Goswami, G. Bhagavannarayana, (Late) Vijay Kumar, S.K. Halder, R.R. Ramanan, Reshmi Mitra, M. Ravikumar, R.V. Ananthamurthy, Ashutosh Choubey, B.P. Singh, G. Srinivas, V.D. Vankar from India and Prof. H.L. Hartnagel, Dr. Peter Thoma, Acad. F.A. Kuznetsov, Yu. V. Shubin, L. Kharchenko and Dr. Vasilev from abroad. The author had the privilege of very long association with Prof. A.R. Verma from whom he has learnt a lot. A part of this work was supported under an Indo-US project coordinated by Department of Science and Technology, New Delhi and a part of the work had been carried out under an Indo-German project coordinated by the Council of Scientific and Industrial Research, New Delhi. The author acknowledges with pleasure the long collaboration with Institute of Inorganic Chemistry, Novosibirsk under ILTP.

The author would like to acknowledge the support received from Indian National Science Academy, New Delhi as INSA Senior Scientist.

References

Authier, Andre, 2001. Dynamical Theory of X-ray Diffraction, Oxford University Press, Oxford.

ASTM, Annual Book of the American Society for Testing Materials Standards, 1980, F1221-80, pp.538-540.

Bullis, W. M., Huff, H. R. R., 1993. ULSI Science and Technology, G. K. Geller, E. Middleworth and K. Hoh (Ed.) The Electrochemical Soc., Pennington, USA,, pp.103.

Bowen, D., Keith, Tanner, B.K., 1998. High Resolution X-ray Diffractometry and Topography, Taylor and Francis Ltd., London EC4A 3DE.

Bergholtz, W. Binns, M.J., Booker, G.R., Hutchison, J.C., Kinder, S. H., Messoloras, S., Newman, R.C., Stewart, R.J., Wilkes, R.J., 1989. Phil. Mag. B, 59, pp.499.

Bhagavannarayana, G., Choubey, A., Shubin, Yu. V., Lal, Krishan, 2005. J. Appl. Cryst. 38, pp.448.

Bhagavannarayana, G., 1994. Ph. D. Thesis, University of Delhi, Delhi.

Bonse, U., Hart, M., 1965. Appl. Phys. Lett. 7, pp. 238.

Batterman, B.W., Cole, H., 1964. Rev. Mod. Phys.36, pp.681.

Claeys, C., Deferm, L., 1996. Solid State Phenomena, 1, pp 47-48..

Choubey, A., 2000. Ph.D. Thesis, University of Delhi, Delhi.

Choubey, A., Bhagavannarayana, G., Shubin, Yu. V., Chakraborty B.R., Lal, Krishan, 2002. Z. Kristallogr., 217, pp. 515.

Dederichs, P. H., 1973. J. Phys. F 3, pp. 471; 1971, Phys. Rev. B4, pp.1041.

Goswami, S. Niranjana N., Lal, Krishan, Wurfl, J., Hartnagel, H.L., 1994. Physics of Semiconductor Devices, Krishan Lal (Ed.), Narosa Publishers, New Delhi, pp. 337.

Goswami, S. Niranjana N., Lal, Krishan, Kothari, R., Bhatnagar, Y. K., 1996. Semiconductor Devices, Krishan Lal (Ed.), Narosa Publishers, New Delhi, pp.382.

Halder, S.K., Kumar, Vijay, Lal, Krishan, Suryanarayana, P., Dixit, B.B., Vyas, P.D., 1994a. Physics of Semiconductor Devices, Krishan Lal (Ed.), Narosa Publisher, New Delhi, pp.337.

Halder, S.K., Kumar, Vijay, Lal, Krishan, Kumar, Vipan, Agnihotri, O.P., 1994b. Physics of Semiconductor Devices, Krishan Lal (Ed.), Narosa Publishers, New Delhi, pp.340.

Kato, N., 1992. J. Acta Cryst. A48, pp.834.

Kato, Y., Fukatsu, S., Shiraki, Y., Vac, J., 1995. Sci Technol, B13, pp.111.

Laudise, R. A, 1972. Analytical Chemistry: Key to Progress in National Progress, W. Meinke and J. K. Taylor (Ed.), NBS Special Publication No. 351, NBS Washington, pp.19.

Lal, Krishan, 1991. Advances in Crystallography and Crystal Growth, Krishan Lal (Ed.), Indian National Science Academy, New Delhi, pp. 125.

Lang, A. R., 1958. J. Appl. Phys. 29, pp. 597; and in Modern Diffraction and Imaging Techniques in Materials Science, S. Amelinckx, R. Gevers, G. Remant and L. Landyut (Eds.), North-Holland, Amsterdam, pp. 407.

Lal, Krishan, 1993. Bull. Mater. Sci. 16, pp. 617.

Lal, Krishan, Singh, Bhanu Pratap, 1977. Solid State Commun. 22, pp.71.

Lal, Krishan, Bhagavannarayana, G., 1989. J. Appl. Cryst. 2, pp.209.

Lal, Krishan, Goswami, S. Niranjana N., Wurfl, J., Hartnagel, H.L., 1990. J. Appl. Phys., 67, pp.4105.

Lal, Krishan, 1998. Proc. Ind. Nat. Sci. Acad. 64A, pp.609.

Laue, M.Von, 1960. Rontegenstrahlininterferennzen, Acad. Verlagsgescllschaft, Frankfurt.

Lal, Krishan, 1995. *Semiconductor Devices*, Krishan Lal (Ed.), Narosa Publishers, New Delhi, pp.243.

Lal, Krishan, Mitra, Reshmi, Srinivas, G., Vankar, V. D., 1996. J. Appl. Cryst. 29, pp.2.

Lal, Krishan, Bhagavannarayana, G., 1989. J. Appl. Phys., 2, pp.209.

Lal, Krishan, Ramanan, R.R., Bhagavannarayana, G., 2000. J. Appl. Cryst., 33, pp.2.

Lal, Krishan, 1989. Prog. in Crystal Growth Charac, 18, pp.227.

Lal, Krishan, Bhagvannarayana, G., Virdi, G. S., 1991. J. Appl. Physics 69, pp. 8092.

Lal, Krishan, Bhagavannarayana, G., Virdi, G. S., 1996. Solid State Phenomena, 47-48, pp. 377.

Lal, Krishan, Singh, Bhanu Pratap, Verma, Ajit Ram, 1979. Acta Cryst. A35, pp. 286.

Lal, Krishan and Singh, Bhanu Pratap, 1980. Acta Cryst. A36, pp.178.

Lal, Krishan and Singh, Bhanu Pratap, 1981. J. Crystal Growth, 54, pp.493.

Metzger, R.A., 1995. Compound Semiconductor, 1, pp. 21.

Murthy, R.V. Anantha, Ravikumar, M., Choubey, A., Lal, Krishan, Kharachenko, Lyudmila, Shlcguel, V., Guerasimov, V., 1999. J. Crystal Growth, 197, pp.865.

Mukhopadhyay, M., Bera, L.K., Ray, S.K., Sahu, S.N., Mehra, B.R., Goswami, N., Lal, Krishan, Maiti, C.K., 1997. IETE Journal of Research, 43, pp.155.

Pinsker, Z. G., 1978. Dynamical Scattering of X-rays in Crystals, Springer-Verlag, Berlin. Rozgonyi, G.A., Ciesielka, T.J., 1973. Rev. Sci. Instrum. 44, pp.1053.

Ramanan, R.R., Bhagavannarayana, G., Lal, Krishan, 1995. J. Crystal Growth, 156, pp. 377.

Ramanan, R.R., Goswami, S. Niranjana N., Lal, Krishan, 1998. Acta Cryst. A54, pp.163.

Ramanan, R. R., 1998. Ph. D. Thesis, University of Delhi, Delhi.

Singh, Bhanu Pratap, 1980. Ph.D Thesis, University of Delhi, Delhi.

Stepanov, S.A., Kondrashkina, E.A., Schmidbauer, M., Koehler, R., Pfeiffer, J. U., Jach, T., Souvorov, A. Yu., 1996. Phys. Rev. B, 54, pp.8150.

Segmuller, A., Noyan, I.C., Speriosu, V.S., 1989. Prog. Crystal Growth Charact. 18, pp.21.

Wooster, W. A., 1962. Diffuse X-ray Reflections from Crystals, Clarendon Press, Oxford.

Zachariasen, W.H., 1945. Theory of X-ray Diffraction in Crystals, Wiley, New York.

Chapter 8

Structural Characterization and Long-Term Reliability of Micro Machined Sensors and High-Resolution X-ray Reflectometry of Solid Surfaces

Krishan Lal and Niranjana Goswami*
National Physical Laboratory, New Delhi – 110 012, India
*E-mail: *klal@mail.nplindia.ernet.in*

ABSTRACT

Micro Electro Mechanical Systems (MEMS) are finding a wide range of applications and therefore, there is a lot of emphasis on R&D in this field at global level. The materials to be employed in fabrication of the systems need to be characterized thoroughly to ensure performance within rigid specifications. Also, as the materials are processed, control on their elemental composition and structure is to be ensured. The fabricated devices have to be free of undue strains and stresses to ensure long-term stability. We have investigated micro-machined sensors based on gallium arsenide single crystals. These have been prepared by combining photolithography, ion implantation and selective etching. The sensors essentially were coiled membranes. High-resolution X-ray diffractometry and topography have been employed successfully for non-destructive evaluation of structural perfection of sensor segments and angular tilts between these. The increase in the stress level in the substrate crystals due to processing for device fabrication has been studied. It was possible to directly photograph the morphological features of the cavity below the membrane sensor. The tilt between adjoining sensor segments could be as low as 15 arc sec. The diffraction curves from individual segments revealed variations in crystalline perfection. The sharpest of curves had a half width of about 70 arc sec, which is quite acceptable. However, the broadest of the diffraction curve showed a lot of strain and had a half width of about 300 arc sec. The scanning electron micrographs did not reveal these details. A new technique for investigation of reflection of highly collimated and monochromated X-ray beams from solid surfaces below critical angle is being established in author's laboratory. It can be used to investigate deposits with thicknesses down to ~ 10 nm.

Keywords: *Micro Electro Mechanical Systems (MEMS), Sensors, Resolution, X-ray.*

Introduction

New exciting developments in the field of sensors and nano-technology offer great possibilities for applications in a wide range of fields. These also pose serious challenges in synthesis, processing, device fabrication and long-term stability and reliability of devices. Deep knowledge about relationship between properties of materials and their basic characteristics namely, composition, purity, structure and defects is of vital importance in all the aspects. Without the control on materials characteristics, it is neither possible to reproduce materials on a large scale nor to produce devices with stringent specifications.

The conventional pressure sensors, *e.g.* the silicon capacitive pressure sensors, the silicon piezoresistive pressure sensors or the piezoelectric pressure sensors have their merits and several applications (Fricke *et al.*, 1993). However, dependence of their sensitivities upon ambient temperature is a drawback. As sensor material GaAs crystals offer operation in a wider temperature range in comparison to silicon due to their larger energy-gap. Further, their radiation hardness due to the direct-gap property and compatibility with opto-electronic integrated circuits are attractive features. The substrate of the sensor can be semi-insulating GaAs, thus giving reduced leakage currents even at higher temperatures. The sensors can be fitted in sensing systems of robots, engines and can be employed in hazardous environments. For higher electromechanical sensitivity, membranes, beams or bridges have formed the basis of sensors for various applications (Wurfl *et al.*, 1992). However, thin membranes suffer from surface loading effects when other layers are deposited on them, which leads to bending of the membrane structures. In author's laboratory detailed structural characterization of pressure sensors based on mechanically stable thick (> 2μm) semi-insulating GaAs membranes has been carried out.

For poly-crystalline as well as single crystal materials, one requires structural information in addition to the elemental composition. A wide variety of techniques is available as well as is under development to accomplish this. These include: a variety of electron optical techniques; high-resolution X-ray diffraction methods; chemical etching and optical systems. In favourable cases, one can use techniques like scanning tunnelling microscopy, which enable atomic level resolution. Most of these techniques except those based on X-ray diffraction are essentially destructive in nature. The high penetrating power of X-rays and their small wavelengths have advantage of non-destructive characterisation as well as gaining structural information at very high resolution.

High–resolution X-ray diffraction techniques have been used for investigation of a variety of crystals, thin films, interfaces and devices including sensors. The five-crystal X-ray diffractometer developed at NPL was employed in (+,–, +) configuration. A highly monochromated and well-collimated Mo $K\alpha_1$ X-ray beam of very narrow lateral width (~ 10 µm) was employed as the exploring beam. This X-ray beam has been of great value in evaluating crystalline perfection in detail. Also, it can be used for reflectometry experiments with new possibilities. This paper reviews some important results.

Experimental Details

Fabrication of Sensors

Figure 8.1 shows a schematic diagram of the free-standing coiled membrane sensors investigated in author's laboratory. Starting wafers were (100) n-type GaAs crystals with n = 3.5 x 10^{17} cm^{-3}. These were implanted with 4 MeV N^{2+} ions at room temperature as shown in Figure 8.2(a) (Fricke *et al.*, 1993; Wurfl *et al.*, 1992; Miao, Ph.D. Thesis, Reihe 9; Miao *et al.*, 1995).The dose level was ~ 10^{15} cm^{-2}. Implantation produced N compensated semi-insulating GaAs surfaces. The projected range of these N^+ ions in GaAs was found to be 3.1 µm. The top ~ 2.5 µm layer was lightly damaged. Below this, a ~ 0.7 µm layer was heavily damaged. Passivation was

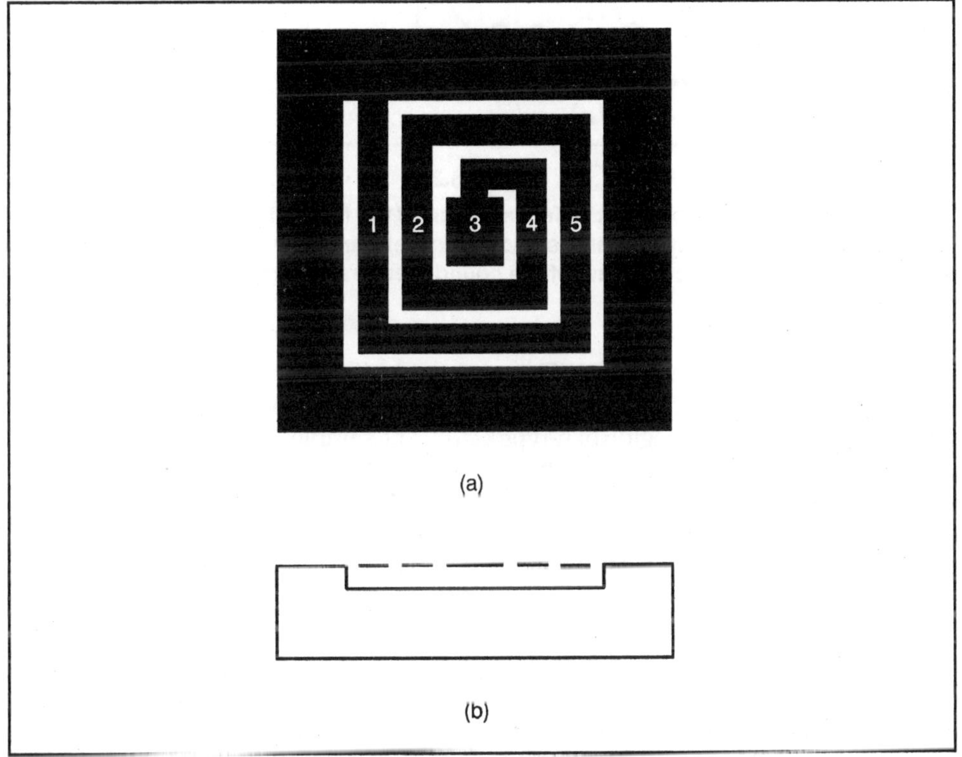

Figure 8.1: A Schematic Diagram of the Free Standing Coiled Membrane Sensor Investigated by High Resolution X-ray Diffraction. (a) top view and (b) side view.

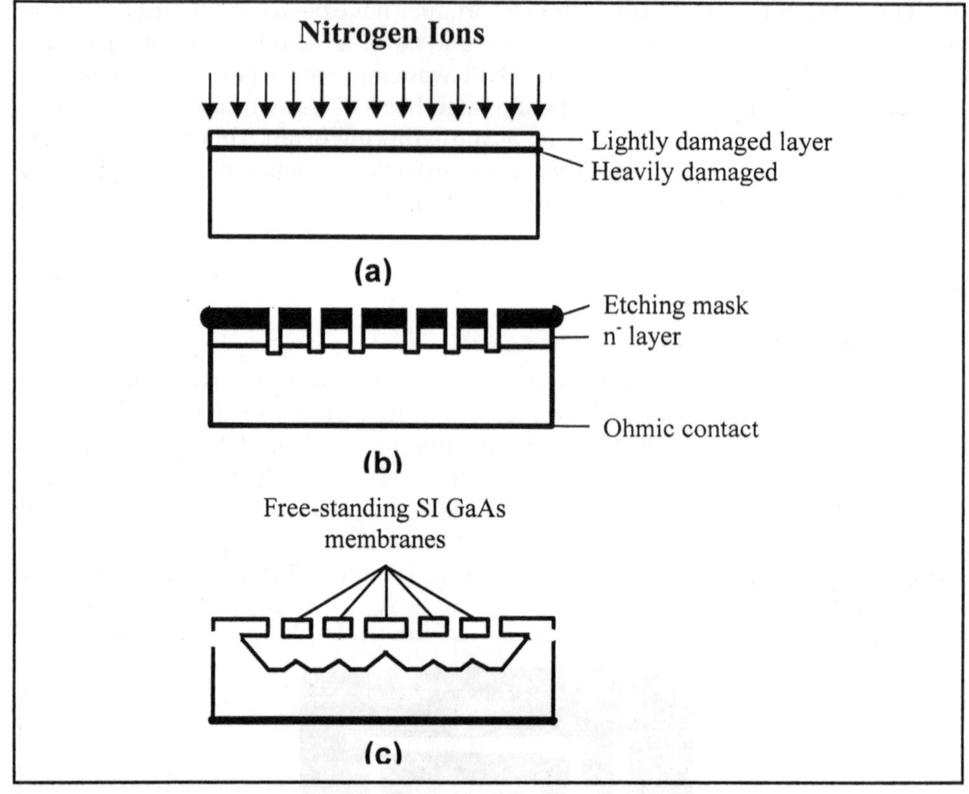

Figure 8.2

(a) Nitrogen implantation of GaAs wafers as a part of the fabrication process of pressure sensors; (b) photolithography and selective ion etching after implantation of GaAs wafers; and (c) schematic diagram of the coiled membrane force sensor fabricated by micro machining.

performed with 200 nm thick PECVD layers of Si_3N_4 films. Annealing at 600° C for 10 minutes in an atmosphere of flowing hydrogen starts compensating the implantation induced lightly damaged region of n-type GaAs. The sample becomes semi-insulating (SI) and shows ideal I-V characteristics after annealing at 800° C. Photolithography was used to define the pattern on photo resist layer, which also acted as an etch mask for subsequent wet chemical etching (Figure 8.2(b)). Selective etching was done by wet chemical etchant, H_3PO_4: H_2O_2: Methanol in volume ratio of 1: 1: 1 for ~ 40 s. Etch depth was found to be ~ 4 μm. To remove highly damaged layer and produce free hanging membrane, pulsed anodic etching was performed. For this, Au-Ge-Ni ohmic contacts were prepared on the back surface of the samples by electron beam evaporation. These contacts were annealed at 450° C for 30 s. For pulsed anodic oxidation, 17 V pulses with time period of 10 ms and duration of 100 μs were employed for a total time of 40 min. Etch rate was observed to be 3 μm/min. This process etched out highly damaged layer and part of the un-implanted n-GaAs and produced freestanding membranes in desired patterns. This process removed ~100 µm of GaAs below ~ 2.5 µm thick membranes (Figure 8.2(c)). In the present study, ~ 5.2 µm long and ~ 70 µm wide coiled membranes with five segments were investigated. The gap between the different segments of the coiled shaped membrane was ~ 30 µm.

Structural Characterization by High-Resolution X-Ray Diffraction Techniques

Figure 8.3 shows a schematic diagram of the Five Crystal X-ray Diffractometer developed in author's laboratory and employed for conducting high-resolution X-ray diffraction experiments (Lal, 1993).

Details of the diffractometer together with some illustrative examples have been described in another paper. It was employed in a three-crystal configuration. A highly monochromated and well-collimated Mo Kα_1 X-ray beam was used as the exploring beam. This was achieved after successive diffractions of X-ray beam emerging from a fine focus X-ray source, which was collimated with help of a long collimator fitted with a fine slit. Two plane silicon crystals of Bonse-Hart type (Bonse and Hart, 1965), arranged in (+,-) configuration were used as monochromators. A Mo Kα_1 X-ray beam of very narrow lateral width (~ 10 µm) was isolated with the help of a fine slit of adjustable width. The specimens formed the third crystal of the diffractometer. Specimens were studied with symmetric (400) as well as with highly asymmetric ('511) diffracting lattice planes in Bragg geometries. The exploring beam height was adjusted according to the membrane size so as to optimize the signal to background ratio. High-resolution X-ray diffraction curves and stationary topographs were recorded after translating the devices across the X-ray beam in steps of 10 µm to cover the entire areas of specimens. High-resolution X-ray diffraction traverse topographs

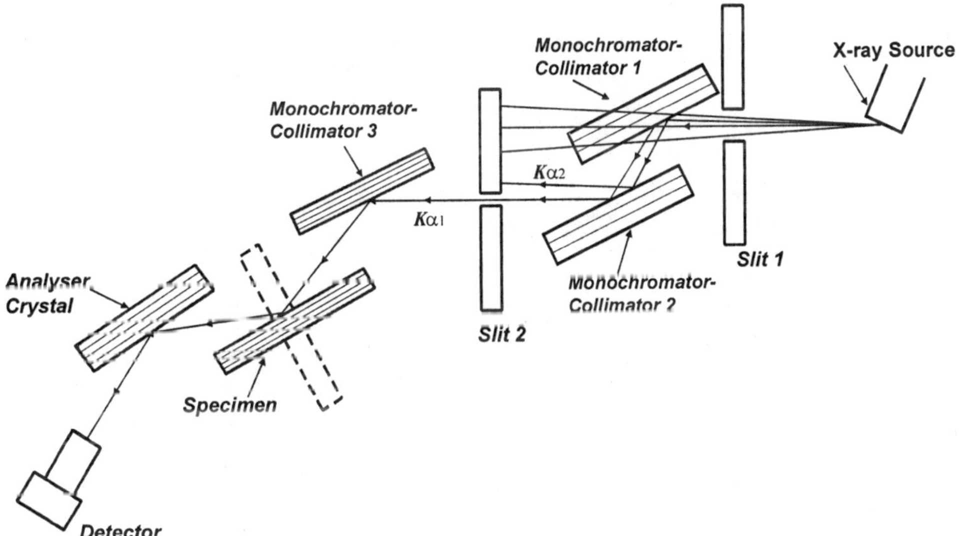

Figure 8.3: A Schematic Line Diagram of the Five Crystal X-Ray Diffractometer Designed, Developed and Fabricated at NPL

were also recorded, covering the membrane region as well as the surrounding substrate region.

Changes in orientations of the diffraction vector were recorded as a function of the linear position of the specimen across the X-ray beam. Results of these curvature measurements were used to quantitatively determine the bending of the specimen and therefore the degree of mechanical stress.

Structural Characterisation of Micro-Machined Sensors

Figure 8.4 shows a SEM micrograph of a typical coil shaped membrane sensor (Lal, 1999). The micrograph shows that the membrane surfaces were quite smooth with well-defined edges. The top surfaces of all the sensor segments appear to be in the same plane and there is no indication of any tilt between these segments. As such, the micrograph of Figure 8.2 indicates good structural quality of the investigated device.

High resolution X-ray diffraction experiments however, revealed interesting details. To begin with gallium arsenide substrate region away from the sensor was investigated. Figure 8.5 shows a typical diffraction curve of the substrate recorded with (400) diffracting planes. This curve is well-defined and shows a single peak (Lal

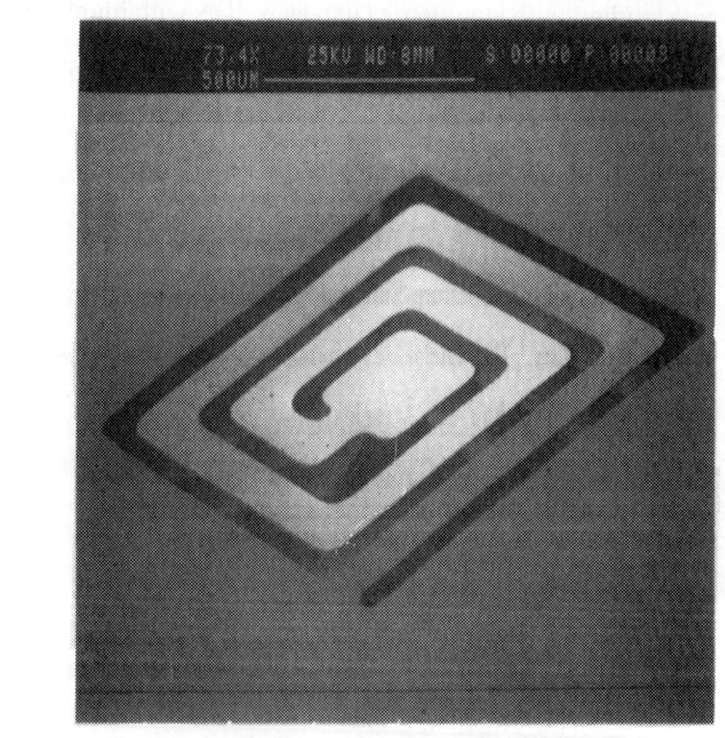

Figure 8.4: An SEM Micrograph of a Typical Coiled Membrane Force Sensor Fabricated by Micro Machining

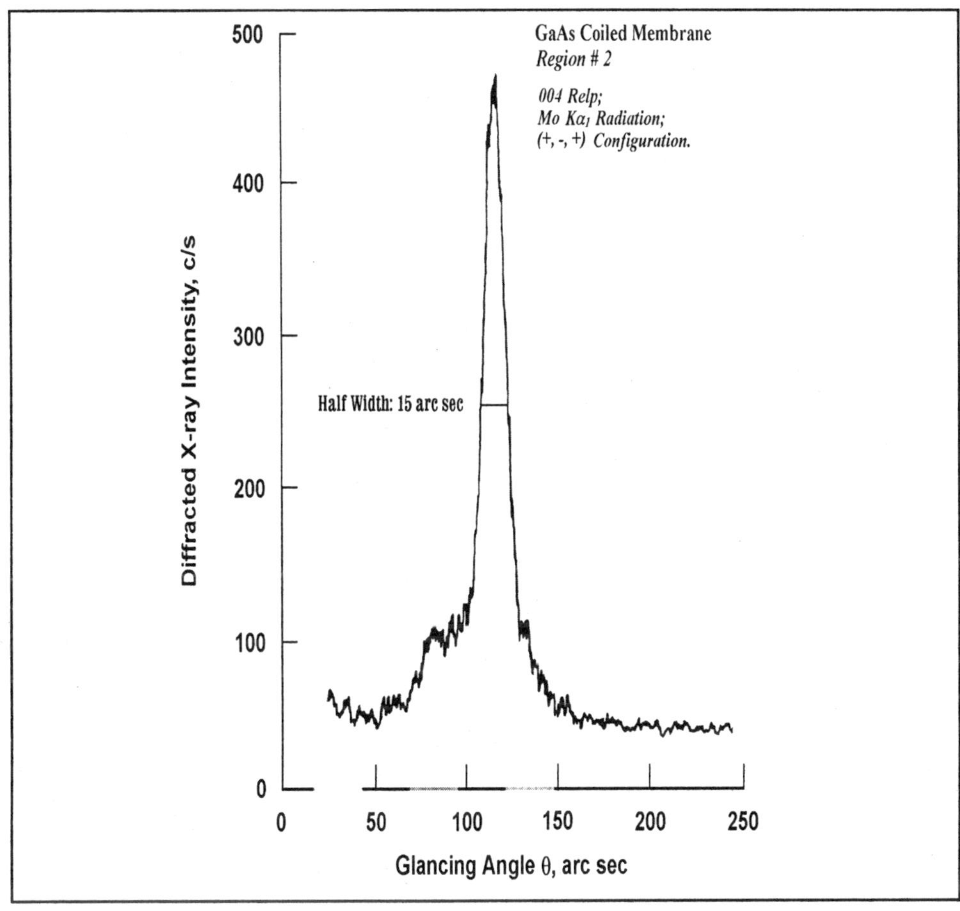

Figure 8.5: A High Resolution X-Ray Diffraction Curve of GaAs Substrate Crystal on which the Coiled Membrane Sensor was Fabricated

et al., 1999). However, it is broad, when compared with the theoretical width of the diffraction curves of gallium arsenide, which is less than ten arc sec. The half widths of the diffraction curves recorded with ('511) diffracting planes from the bulk GaAs crystals surrounding the sensor region were ~ 26 arc sec. A fraction of the broadening of the diffraction curve is due to dispersion as a result of the different lattice spacings of the diffracting planes of the silicon monochromator crystals and those of the ('511) diffracting planes of the GaAs specimens. The substantial bending of specimen, which was found from curvature measurement studies, also leads to broadening of diffraction curves. The structural damage due to processing of the wafers was responsible for the rest of the broadening. Keeping in view all the factors responsible for broadening of diffraction curves, it is concluded that the starting wafers had good degree of crystalline perfection. High-resolution diffraction curves of device quality GaAs crystals have typical half widths of about 10 arc sec or less. However, keeping in view the fact that the crystal was subjected to processing steps like ion implantation, this is a reasonably good diffraction curve.

The sensor-region was investigated in detail by high resolution X-ray diffraction. The sensor was scanned stepwise across the exploring X-ray beam to investigate its different segments.

Each time it was moved by a step of 10 μm. Figure 8.6 shows a typical diffraction curve recorded when the exploring beam dimensions and sensor position were adjusted for the central part being explored (Lal *et al.*, 1999). Well-defined peaks are observed due to diffraction from three different segments of the sensor. Similar curves were recorded to cover the entire sensor after linearly moving the sensor in a step wise manner, as mentioned above. A comparison and study of all the diffraction curves gives detailed structural information about the sensor. There are complicating factors while investigating such devices. The diffracted beam can be from the cavity below the sensor as well as from the sensor segments themselves. By detailed experiments the diffraction curves of the sensor segments had to be isolated and analysed. The half widths of the diffraction curves of sensor-segments varied in a broad range. One of the curves in Figure 8.6 has a half width of 72 arc sec. The

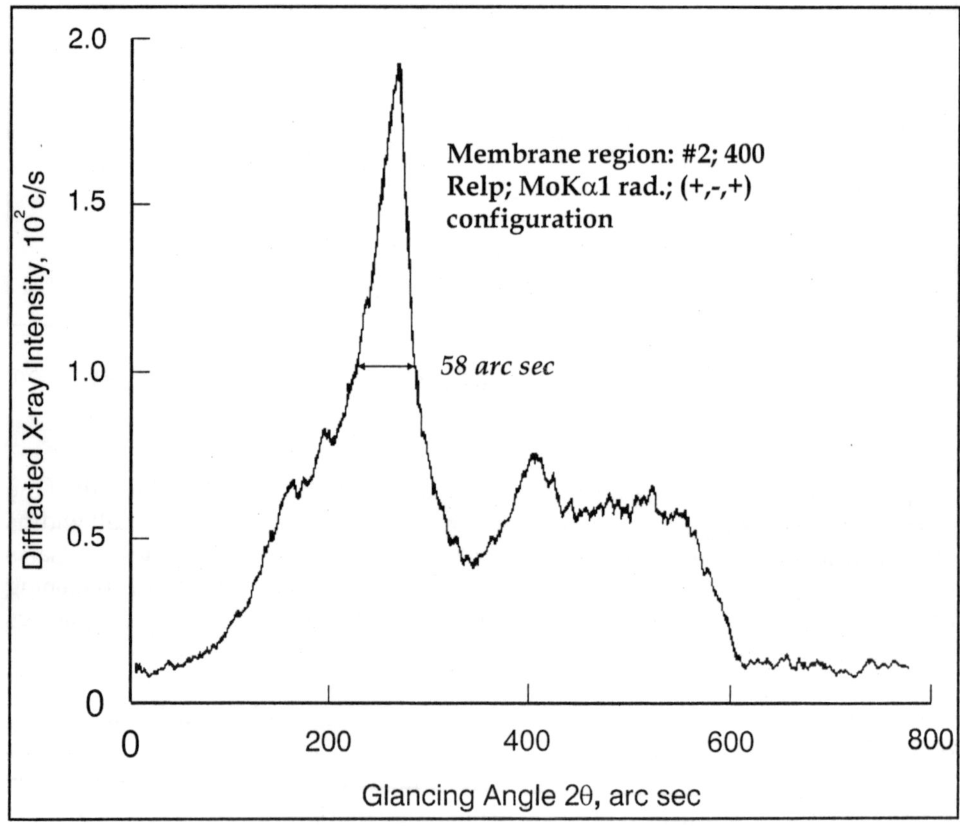

Figure 8.6: A Typical Diffraction Curve of a Micromachined Membrane Sensor Recorded from a Specific Region (#2) with MoKα1 Radiation. Diffraction peaks due to different segments can be seen in the figure.

presence of more than one peak in the diffraction curves clearly shows that there are different segments of the sensor have angular mis-orientations with respect to each other.

The values of the angular tilts between the adjoining segments can be determined from the angular separation between the peaks. When experimental data about tilts obtained from all the diffraction curves was analysed a total picture about tilts emerged. Table 8.1 lists all the tilts observed between different segments. These lie in the range: 15 arc sec to 212 arc sec (Lal *et al.,* 1999). It may be mentioned that investigation by scanning electron microscopy did not reveal any tilts between the sensor segments.

Table 8.1: Tilt Angle between Various Segments of the Coiled Membrane Sensor

Membrane Segment No	*Tilt with Respect to Left Hand Side of the Substrate Arc Sec*	*Tilt Angle Between Adjoining Segments Arc Sec*
#1	33	29
#2	62	15
#3	77	212
#4	289	95
#5	384	

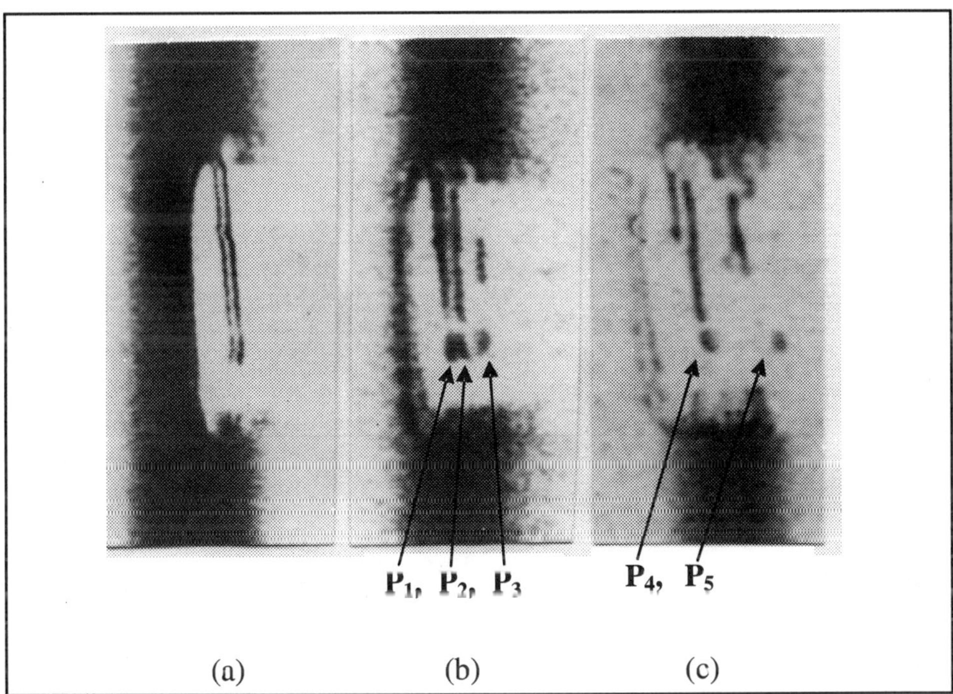

Figure 8.7: A Set of High Resolution X-Ray Diffraction Topographs Recorded from Three Different Regions of the Coiled Membrane Sensor. Images of diffracted beams from the sensor segments are indicated by arrows.

X-ray diffraction topographic examination of the sensors in symmetrical as well as asymmetrical reflection geometries did yield interesting results. Firstly, it was observed that the entire set of segments were not diffracting at any angular orientation of the sensor, due to tilts between segments as demonstrated by diffractometric evaluation. In general, a fraction of segment area only satisfied the diffraction condition. Figure 8.7 shows a set of topographs revealing these tilts (Lal *et al.*, 1999).

As mentioned above the X-ray beam can reach the cavity below the sensor and by adjusting experimental conditions one could record diffracted images from the cavity, which reveal its morphology in a non-destructive manner. This investigation revealed interesting results. Figure 8.8 shows pictorially the experimental arrangement employed for these experiments. On the right hand side an image of the diffracted beam (topograph) has been shown (Figure 8.8) (Lal *et al.*, 1999). This topograph shows the following interesting features. The profile of the cavity shows that it is not flat and its depth was found to be ~ 90 μm. The unevenness of the cavity was determined as ± 4 μm. It was interesting to note the image of the over-hanging part of the sensor on the cavity as clearly seen in the image of Figure 8.8. Further, this image shows the sides of the cavity are tapered and make an angle of about 57° with the top surface. This is due to the preferential etching along (111) planes. In general, the sidewalls made angles lying in the range of 55° to 60°.

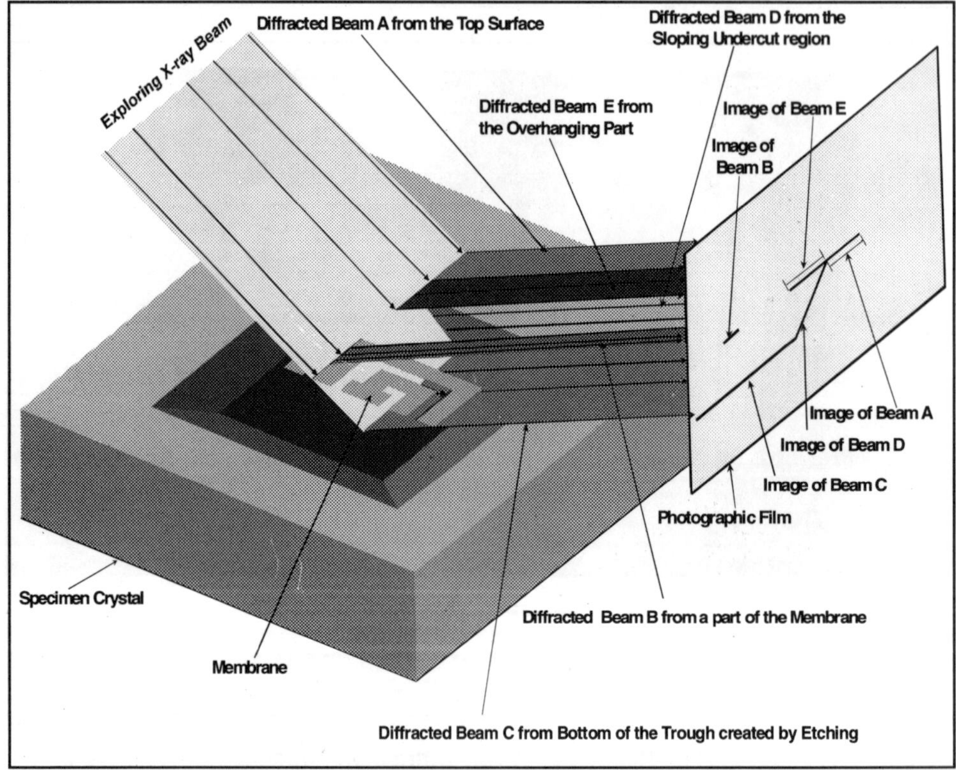

Figure 8.8: A Perspective View of Diffraction of X-Rays from the Coiled Membrane Sensor

Figure 8.9 shows a profile of the sensor as revealed by high resolution X-ray diffraction study (Lal *et al.*, 1999). The in-built stress has bent the entire sensor. The adjoining segments of sensor are tilted in respect to each other. The shape of the cavity is also depicted schematically. The information on structural characterization has served as a useful feedback for improving design of the micro-machine sensors. Also, it helps in reducing the stress levels so that long-terms stability of the device can be enhanced.

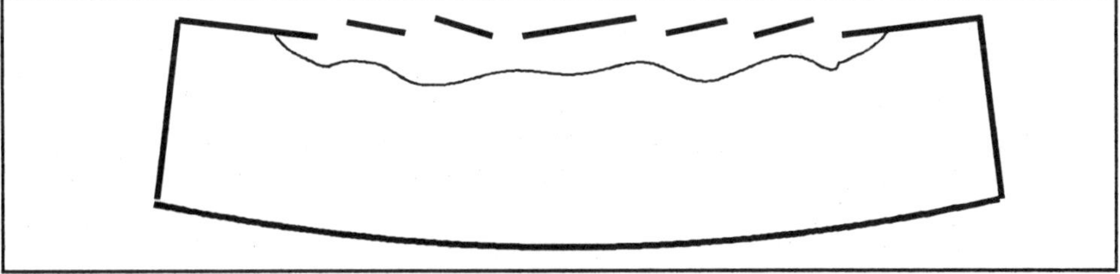

Figure 8.9: A Schematic Diagram of the Final Shape of the Sensor and the Crystal on which it has been Fabricated

High Resolution X-Ray Reflectometry for Characterization of Solid Surfaces, Interfaces and Thin Deposits

It is well known that the refractive index of solids is slightly less than unity at X-ray frequencies.

As a consequence specular reflection can be observed when an X-ray beam is incident at any glancing angle that is ≥ the critical angle w_c. The critical angle is a function of the refractive index of the material. The technique of X-ray reflectometry is now being extensively used for investigation of solid surfaces, thin deposits and interfaces. Information about important parameters like composition, film thickness and surface roughness can be obtained. A new technique for study of X-ray reflections from solid surfaces is being developed in author's laboratory. A brief description of this new approach and an illustrative result are briefly described below.

Figure 8.10 shows a schematic diagram of the experimental set-up being developed for high resolution X-ray reflection by author and co-workers. X-ray beam from a

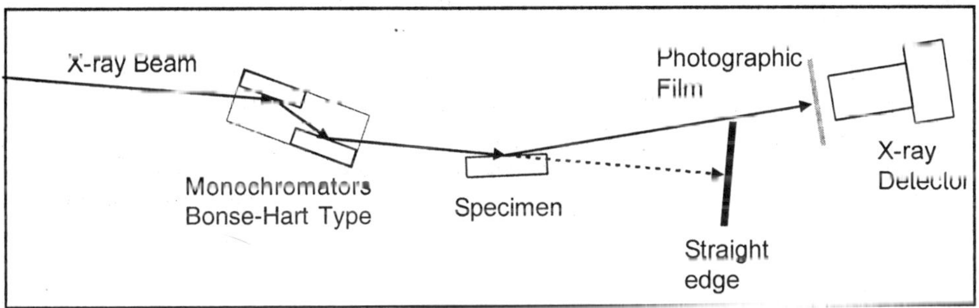

Figure 8.10: A Schematic Diagram of the High Resolution X-Ray Reflectometry Setup being Developed at NPL

high brilliance X-ray generator is collimated by a long collimator, which is fitted with a fine slit. It is then diffracted from two plane silicon monochramators of Bonse-Hart type (Bonse and Hart, 1965) as shown (Figure 8.10). The diffracted beam is a highly collimated and monochromated Kα_1 beam, Mo Kα_1 beam in these experiments. This is employed as the exploring X-ray beam. It combines the high level of collimation and monochromaticity with very small width in the plane of reflection (~100 μm or less). The specimen is mounted on a special turntable, which is used to precisely position and orient it for specular reflection of X-ray beam. High resolution X-ray reflection curves are recorded by measuring the intensity of the reflected X-ray beam as a function of the glancing angle. The intensity of the reflected beam is measured by employing a scintillation counter. A straight edge is inserted to block the residual exploring beam and allow only the reflected beam to reach the detector or the photographic film. The images of the reflected beam are recorded on high resolution X-ray films. These directly recorded images give very interesting details. The analysis of the images has been employed in identification of the chemical species present on the specimen and for determination of their composition.

Figure 8.11 shows a set of two reflection curves of silicon. The outer one is a theoretical curve obtained by using the electro-magnetic theory of reflection. The inner curve is that obtained with the experimental set up described above (Figure 8.10). The parameter β and δ have been optimized for best match between the theoretical and the experimental curves. It can be seen that the two curves have reasonably good agreement.

Several experiments have been performed with the set up described above for study of X-ray reflections from crystal surfaces. Surfaces of materials like silicon, gallium arsenide and garnet have been investigated. Figure 8.12 shows an image of the reflected beam obtained from a silicon crystal surface. Several parallel lines

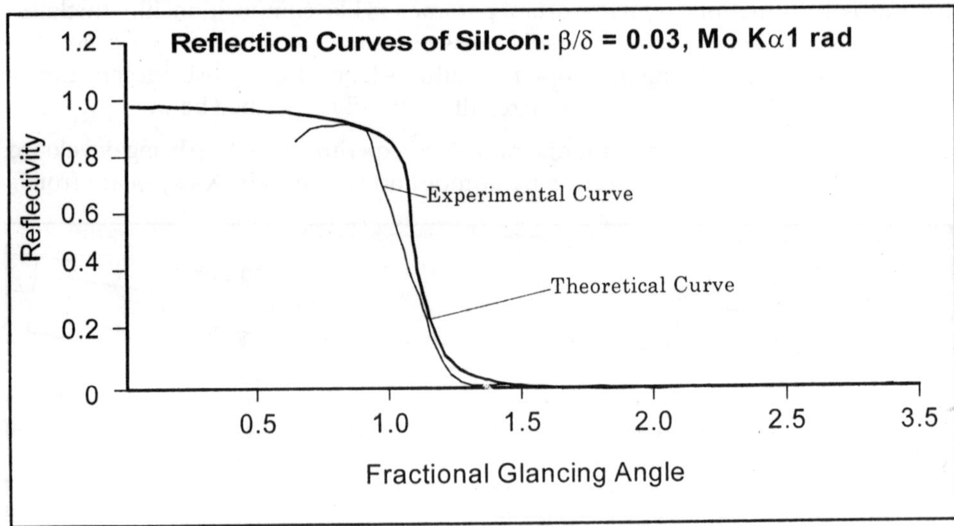

Figure 8.11: An Experimental and a Theoretical Reflection Curves of silicon. The theoretical curve computed with β/δ =0.03 (shown here) gave satisfactory fit.

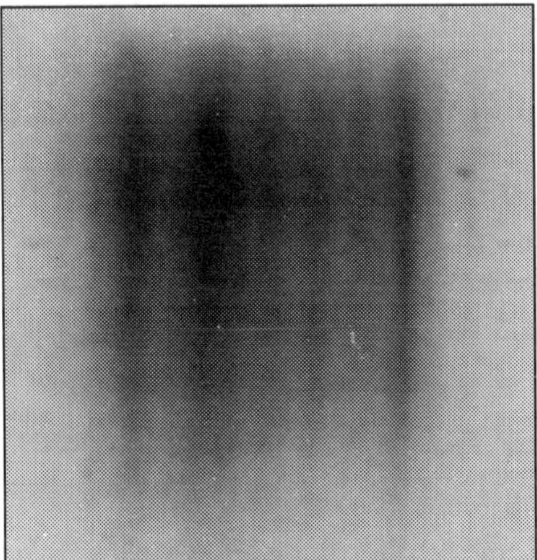

Figure 8.12: A Typical Reflectogram (image of the reflected beam) of a Silicon Crystal Surface. Different lines in this picture represent species present on the specimen surface with their specific values of refractive index.

observed in this image indicate presence of different layers of varying refractive indices on the surface of the specimen crystal. From the analysis of this pattern, the value of x in SiO_x has been determined. These results have demonstrated the presence of several types of oxides on the silicon surface, as expected. The strength of the new technique lies in the direct imaging capability. It is possible to detect different species on solid surfaces and to determine their chemical identity.

Conclusion

It has been demonstrated that the stable, single crystalline thick GaAs membranes in complicated shapes could be reproducibly synthesized. Non-destructive high-resolution X-ray diffraction techniques could be successfully employed for determining the structural quality and level of biaxial stress of the complicated micro-machined structures. This study was successful in qualitatively determining the tilt between adjoining segments of the sensors. The results of the structural characterization can help in improving the long time stability of devices. The new high resolution X-ray reflectometry technique being developed at NPL has high potential of revealing chemical species on solid surfaces. The possibility of direct imaging of reflected beams is a unique feature of the technique.

Acknowledgement

A part of this work was supported under an Indo-German collaborative project. It is a pleasure to acknowledge collaboration with Professor H.L. Hartnagel's group at Technical University Darmstadt, Darmstadt, Germany

References

Bonse, U., Hart, M., 1965. Appl. Phys. Lett. 7, pp.238.

Baake, O., Oksuzoglu, R. M., Flege, S., Hoffmann, P. S., Gottschalk, S., Fuess, H., Ortner, H. M., 2006. Mater. Charact. 57, pp.12-16.

Chason, E., Mayer, T. M., 1997. Crit. Rev. Solid State Mater. Sci. 22, pp.1-67.

Fricke, K., Wurfl, J., Miao, J., Dehe, A., Ruck, D., Hartnagel, H.L., Micromech, J., 1993. Microeng, 3, pp.131-134.

Krishan Lal, Goswami, S. Niranjana N., Miao, J., Hartnagel, H.L., 1999. Proc SPIE Vol 3903, pp. 77.

Lal, Krishan, Goswami, S. Niranjana N., Miao, J., Hartnagel, H.L., 1999. J. Appl. Cryst. 31, pp. 60-64.

Lal, Krishan, 1993. Bull. Mater. Sci. 16, pp.617-642.

Miao, J., Ph.D. Thesis, Reihe 9; Elektronik Nr.232 VDI Verlag, Technical University Darmstadt, D-64283, Darmstadt, Germany.

Miao, J., Hartnagel, H.L., Weiss, B.L. and Wilson, R.J., 1995. Electron. Lett. 31, pp.1047-1049.

Parrat, L. G., 1954. Phys. Rev. 95, pp.359-369.

Wurfl, J., Miao, J., Ruck, D., Hartnagel, H.L., 1992. J. Appl. Phys., 72, pp.2700-2704.

Chapter 9

Scanning Probe Microscopy for Biological Imaging

Gerald Kada[1] and David P. Allison[2]
[1]*Nanotechnology Measurements Division, Agilent Technologies, Mooslackengasse 17, 1190 Vienna, Austria*
E-mail: gerald_kada@agilent.com
[2]*Department of Biochemistry and Cellular and Molecular Biology The University of Tennessee, Knoxville, Tennessee, 37932, USA*

ABSTRACT

This paper gives details about the Scanning Tunneling Microscope (STM) and Atomic Force Microscope (AFM). The STM was able to resolve atomic structure by raster scanning a conductive tip, mounted on a piezoelectric ceramic, over a conductive sample. Although there were several critical differences, one might consider that the invention of AFM was an extension of STM. Unlike the STM, the AFM was able to image insulating material and results were easier to obtain. Commercial scanning probe instruments typically fall into two categories, sample scanning or device scanning. A unique attribute of the scanning probe microscope is its ability to measure forces. Molecular Recognition Force Microscopy (MRFM) is another area where AFM will play a significant role in future biological research.

Keywords: *Scanning tunneling microscope, Atomic force microscope, Commercial scanning probe instruments, Atomic structures, STM images, Biomolecules, Instrumentation, Biological research.*

Introduction

About 25 years ago Gerd Binnig and Heinrich Rohrer published the first images taken with a Scanning Probe Microscope (SPM) (Binnig and Rohrer, 1982). Their new invention, the Scanning Tunneling Microscope (STM), was able to resolve atomic structure by raster scanning a conductive tip, mounted on a piezoelectric ceramic, over a conductive sample. The incredible resolution of this new microscope was met with great enthusiasm within the international scientific community and laboratories throughout the world began using this instrument. The obvious implications of a high-resolution microscope that could image in air was not lost on biologists who teamed with physicists to develop techniques for imaging biological samples. Although STM images of bio-molecules were published, STM images were difficult to reproduce and this frustrated attempts to use the instrument as a tool in biological research. Nevertheless, STM was responsible for establishing a fresh new focus on microscopy. Consequently, research groups throughout the world embraced the next generation of SPM instruments which they envisioned would impact biological research. This new instrument was the Atomic Force Microscope (AFM), that was invented in 1986 (Binnig *et al.*, 1986) and became commercially available in 1989. A number of review articles and books offer a comprehensive review of either AFM instrumentation or biological applications and can be used as starting points for people who are interested (Bustamante and Rivetti, 1996; Hansma *et al.*, 1997; Morris and Kirby, 1999; Watanabe and Sakurai, 2000; Bonnell, 2000). This article focuses on the evolution of the AFM as a valuable tool for biological research in the twenty-first century.

One might consider the AFM as an extension of the STM, but it had several critical differences. For example, instead of having an electronic probe on the end of the scanning piezoelectric ceramic to electronically map a surface, the AFM used a sensitive cantilever with a sharpened tip. A computer image of the sample was created by raster scanning the cantilever over the surface, acquiring a series of data points in the X direction, and making an equal number of scans in the Y direction. Height information was obtained by focusing a laser beam on the back of the cantilever that was reflected into a position sensitive diode. As the cantilever moved over the surface and responded to changes in surface topography, a feedback loop applied a voltage to the Z domain of the piezoelectric ceramic. The piezoelectric element moved the cantilever up or down in order to maintain a fixed position for the laser beam striking the photodiode. This compensating voltage served as the height information for the topographic image.

Unlike the STM, the AFM was able to image insulating material and results were easier to obtain. AFM images of proteins (Quist *et al.*, 1995; Radmacher *et al.*, 1994; Hallett *et al.*, 1995) DNA (Bustamante *et al.*, 1992; Thundat *et al.*, 1992; Hansma *et al.*, 1992; Lindsay *et al.*, 1992) and even images of whole living cells (Henderson *et al.*, 1992; Radmacher *et al.*, 1992) began to appear in the literature as scientists embraced this new instrument.

Future Biological Research

A remarkable characteristic of biological systems is the size scale upon which they function. All living species, regardless of size, are organized down to the molecular scale. Genetically encoded instructions, thousands of different gene-based products, various small molecules and metabolic reactants and products are arranged and manipulated within a microscopically small volume. With genomic sequence in hand new biological initiatives are focusing on better understanding the role that genes and their coded proteins play in the structural-functional organization of cells.

Unraveling these complex systems will require experimental techniques that can identify, localize, and quantify interactions between component molecules. New and improved instrumentation will be required to meet these new research challenges. SPM techniques can assume a benchmark role in the post-genomics era by providing absolute scale resolution and new approaches to screening. Of course, these attributes are not limited to biological uses, applications in material and chemical screening would also be enabled by SPM based techniques.

Evolution of the AFM

Imaging with the AFM

As already outlined above, the AFM operates by monitoring the position of a sharpened tip, supported on a microcantilever as the AFM tip is raster-scanned over a surface. The first instruments that were developed imaged in contact mode in which a sharp AFM tip physically interacted with the sample (Binnig *et al.*, 1986). This interaction with the sample created a significant amount of lateral force and in many cases it required immobilization techniques for holding biomolecules to surfaces. For example, DNA molecules were immobilized on mica surfaces by introducing divalent cations such as magnesium, cobalt, or nickel, to the mounting media (Bustamante *et al.*, 1992; Thundat *et al.*, 1992). Alternatively, by reacting silane compounds with positive head groups, such as aminopropyl-triethoxysilane (APTES), to mica surfaces DNA could be attracted and immobilized for AFM imaging (Lindsay *et al.*, 1992; Lyubchenko *et al.*, 1992).

Techniques for immobilizing biomolecules to surfaces are an ongoing challenge that is beyond the scope of this paper. However, there are many reviews and book chapters that deal with this issue and serve as excellent sources for the interested reader (Bustamante and Rivetti, 1996; Hansma *et al.*, 1997; Morris *et al.*, 1999; Watanabe and Sakurai, 2000; Bonnell, 2000).

Non-Contact AFM

There are two types of non-contact imaging modes, acoustic drive and magnetic ac mode. In acoustic drive, contact with the sample is intermittent. The cantilever is oscillated by applying a voltage to a piezoelectric actuator (PZT) that generates sound waves in the cantilever holder that causes the cantilever to oscillate at its resonant frequency. As the cantilever tip is brought proximal to the sample surface, van der Waals and other surface forces interact with the oscillating cantilever causing a dampening of the oscillation. The reduction of the oscillation amplitude is recognized by the instrument software to detect the surface and construct sample topography (Hansma *et al.*, 1994). This type of non-contact AFM was pioneered by Digital Instruments (Santa Barbara, CA) and is known as tapping mode AFM.

The second mode of operation is truly a non-contact mode of operation, called MAC Mode, pioneered by Molecular Imaging (Tempe, AZ; now Agilent Technologies Inc., Chandler, AZ). In this mode of operation a cantilever coated with a magnetic material is excited to resonance by placing a solenoid proximal to the cantilever and applying an ac voltage to the solenoid creating an electromagnet. Reduction of the oscillation amplitude of the cantilever as it approaches the surface is used to detect proximity of the cantilever tip to the surface and subsequently produce a topographic image (Lantz *et al.*, 1994; Han *et al.*, 1996). In MAC Mode imaging, deformation of the sample is not a problem and image contrast can be extremely good. Also, when imaging in liquid, since only the cantilever is oscillated and not the entire cantilever holder assembly, as in tapping mode, there is less disturbance due to sound waves in the liquid. This results in superior image quality and has obvious utility for meeting the challenges that will be made for instrumentation for contemporary dynamic biological imaging.

Sample Scanning Vs. Cantilever Scanning

In general, commercial scanning probe instruments typically fall into two categories, sample scanning or device scanning. Operating as an AFM, sample scanning keeps the cantilever stationary and moves the sample beneath the cantilever tip during an imaging process. Therefore, the scanning mechanism is always associated with the platform that the sample is mounted on. This creates a major problem in that the area beneath the cantilever is severely limited for the adaptation of a variety of applications for mounting the sample. For example, placing a heating-cooling stage directly on the piezoelectric scanning mechanism can be problematic. Another complication exists with this configuration when imaging in a liquid environment since any leakage from the liquid operation posts a significant risk for damaging the scanning mechanism due to the high bias voltage required to operate the piezoelectric scanner.

The second scanning category is device scanning, commonly referred to as top-down scanning, where, with AFM, the cantilever is mounted on the scanning mechanism and brought down onto the sample. This avoids the problems, just described, with sample scanning and also offers significant options for placing the AFM onto a fluorescent microscope or, due to the large space beneath the scanner, flexibility for implementing alternative sample stage configurations. This would include such options as heating-cooling stages, and a movable stage beneath the scanning mechanism for positioning microarrays for AFM imaging and analysis. However, since most of the AFM instruments use a detection system based on laser beam deflection, cantilever scanning requires a laser tracking system to keep the laser spot on the cantilever during imaging. Such a system is usually more complicated and requires great precision in assembly. This is because any relative movement between the laser spot and cantilever, due to imperfection of beam tracking will cause undesired effects in the image. For example, an abnormal bow effect that makes the image appear concave instead of flat. Also, characteristic waving in the image background structures may appear. This problem has been solved by Agilent Technologies with their new instrument, the 5500 AFM, by a patented beam tracking system that maintains the laser spot in a fixed position on the cantilever during imaging.

AFM as a Research Tool in Biology

As both AFM instrumentation and biological sample preparation matured it became relatively routine to image a variety of biomolecules and even living cells. Getting high-resolution images of samples like the bacterial S-layer protein shown in Figure 9.1, imaged in MAC Mode both in air and liquid became commonplace. Imaging mammalian cells does not require techniques for anchoring cells to surfaces and a number of laboratories have successfully imaged living cells in liquid environment. However, imaging bacterial cells requires immobilization techniques, like gelatin surface on mica (Doktycz *et al.*, 2003) for imaging live bacterial cells in liquid as shown in Figure 9.2.

Most of the early AFM papers published on biological imaging and even papers published today will have in their introduction a statement proclaiming that the AFM is a instrument capable of imaging in any environment including liquids making it especially attractive for biological applications. This implication is becoming a reality today as the AFM takes its rightful place as an imaging tool for investigating

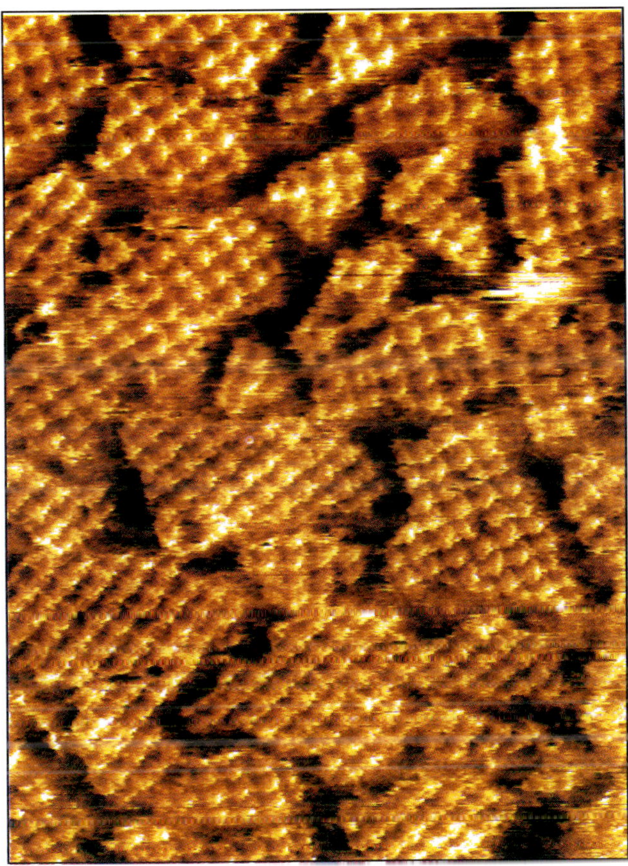

Figure 9.1: High-resolution AFM Image of S-layer Proteins Isolated from Bacteria and Imaged by MAC Mode in Liquid. Image size 500 nm. 5 nm full height scale.

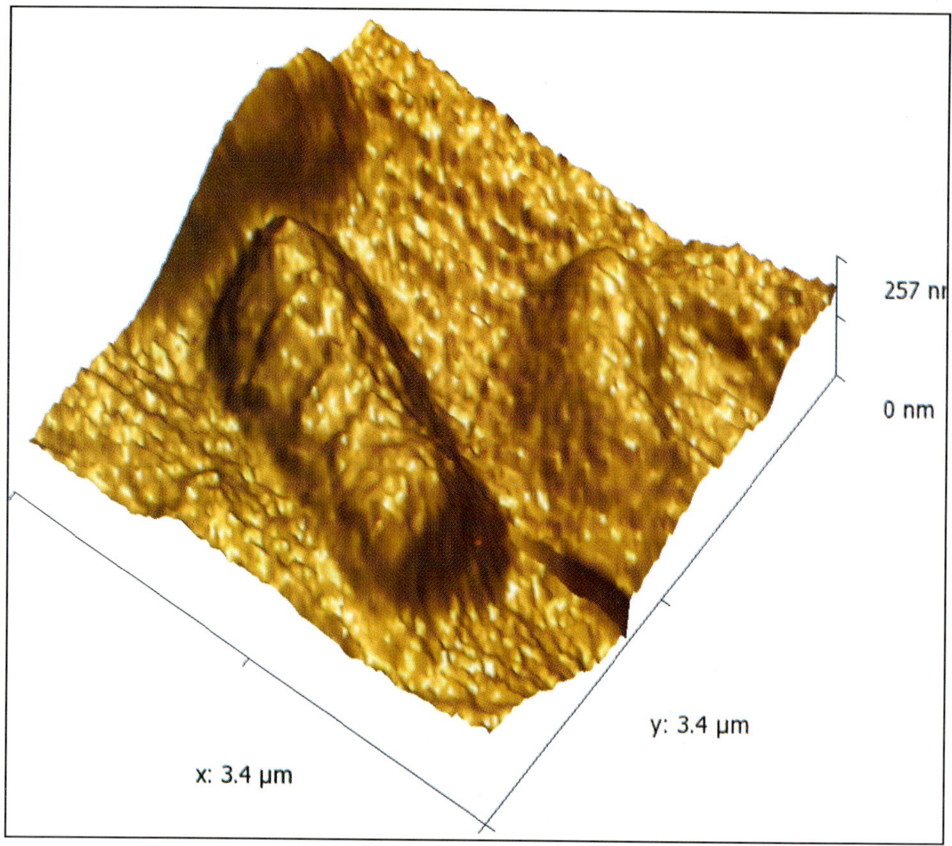

Figure 9.2: Living Bacteria Cells in Buffer Solution Imaged with MAC Mode AFM. Image size 3.4 μm. 300 nm full height scale.

real biological problems both in air and especially in liquids. For example, restriction mapping DNA molecules by AFM imaging (Allison *et al.*, 1996, 1997) was accomplished by binding a mutant restriction enzyme (EcoRI) that would site specifically bind to DNA molecules but would not cleave the DNA. Images taken in air, as in Figure 9.3, showed that this technique could be effectively used to map cosmid sized (30-50 kb) DNA molecules. The ability to image DNA in liquid environments (Hansma and Laney, 1996; Thompson *et al.*, 1996) has led to a number of dynamic studies. By limiting the concentration of nucleotide triphosphate (NTP) it was possible to image transcription as the double-stranded DNA molecule moved through the RNA polymerase molecule and, after initiating binding of the RNA to mica, image the RNA transcription product (Kasas *et al.*, 1997). Another dynamic study done in liquid showed that changes in protein shape can be observed by AFM during enzymatic activity (Radmacher *et al.*, 1994). An especially elegant study from Lindsay's laboratory (Wang *et al.*, 2002) shows changes in a nucleosome DNA complex due to changes in salt concentration. This dynamic study was accomplished

Figure 9.3: AFM Image of an Entire Mouse Cosmid (35 kb) Imaged in Air. The dark spots are EcoRI endonuclease proteins site-specifically bound to the DNA molecule. This illustrates an application for restriction mapping DNA molecules by AFM imaging. Image size 500 nm.

by binding a DNA polynucleosome array to a mica surface, treated with APTES that was activated by glutaraldehyde, and imaging in MAC Mode, as shown in Figure 9.4, in a flow through cell. At low salt concentrations differences in size of nucleosomes was observed while at high salt concentrations the DNA was removed from the

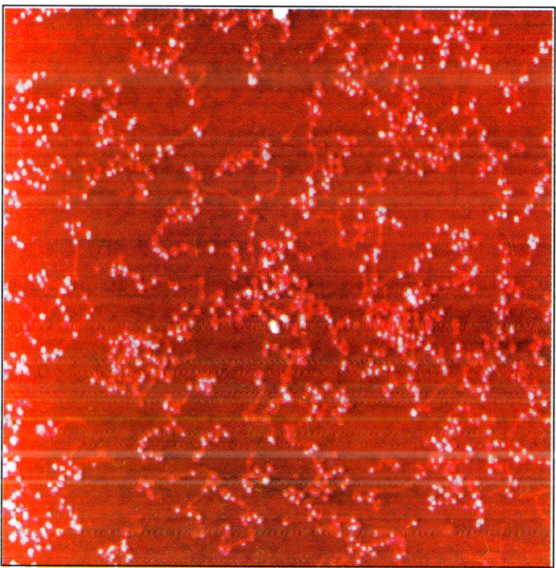

Figure 9.4: MAC Mode Image of a DNA Fragment Coated with Nucleosomes Imaged in a Liquid Cell. Differences in the size of nucleosomes could be produced by changes in the buffer salt concentration. Image size 1.5μm.

nucleosomes. Dynamic studies have not been confined to biomolecules, structural changes in mammalian cells have also been documented, such as the destruction of the actin network in fibroblasts by cytochalasin B (Rotsch and Radmacher, 2000).

AFM Force Spectroscopy

A unique attribute of the scanning probe microscope is its ability to measure forces. As the cantilever tip is approached to the surface and interaction occurs force sensitivity results are obtained by measuring sub-Angstrom deflections of the weak spring (cantilever) as the cantilever is retracted from the surface. This combination can quantify molecular scale forces as small as a single hydrogen bond or weak van der Waals interactions (Hoh *et al.*, 1992). This extreme sensitivity of the cantilever can be exploited to measure forces both within and between biomolecules. Forces required to rupture bonds between complementary oligonucleotides have been accomplished by attaching one single-stranded DNA molecule to the cantilever tip and its complement to the surface and after annealing retracting the cantilever and pulling the molecules apart (Lee *et al.*, 1994a; Boland and Ratner, 1995). Experiments similar to these have been done by attaching biotin to the cantilever and either avidin (Florin *et al.*, 1994; Ludwig *et al.*, 1994) or streptavidin (Lee *et al.*, 1994b) to the surface and after interaction measuring the forces required to pull the molecules apart. Another report describes attaching a single eukaryotic cell to the cantilever and reacting it with a cell growing in a petridish and measuring the force required to break the interaction of the adhesion protein CsA between cells (Benoit *et al.*, 2000). The forces reported to rupture interactions between molecules have been reasonably consistent when reported by different laboratories. For example, a gold coated tip and a gold coated surface were modified with 11-mercaptoundecanoic acid by reacting the thiol groups with the gold and allowing the carboxylic acid groups to interact. The force required to break a single hydrogen bond was found to be 16.6 pN (Han *et al.*, 1995). In a similar experiment where hydrogen bonds were allowed to interact the rupture force was found to be 12 pN (Hoh *et al.*, 1992).

A force extension or force distance curve is generated by approaching a surface with the cantilever tip, touching the surface, and retracting the tip. If a biomolecule becomes attached between the tip and the surface the distance traveled to retract the tip and forces required to extend the biomolecule can be determined. The mechanical proteins titin (Rief *et al.*, 1997; Marszalek *et al.*, 1999; Tskhovrebova *et al.*, 2000) and tenascin (Oberhauser *et al.*, 1998) have been extensively studied with this type of single molecule force spectroscopy. It should be pointed out that in the literature this type of analysis is often called either dynamic force microscopy (DFM) or dynamic force spectroscopy (DFS). In a typical experiment the protein is absorbed on a gold coated mica or glass surface, the cantilever tip is brought into contact with a force of a 1-2 nN for a couple of seconds, and retracted. If a protein is attached a typical saw tooth pattern will be observed in the force distance curve. A typical result for titin is shown in Figure 9.5, where each minimum of the cantilever deflection indicates a force maximum where an individual domain of the titin molecule unraveled.

The distances between the minimum peaks correspond to the length of a domain. For titin reported maximum force peaks varied between 150 to 300 pN with a

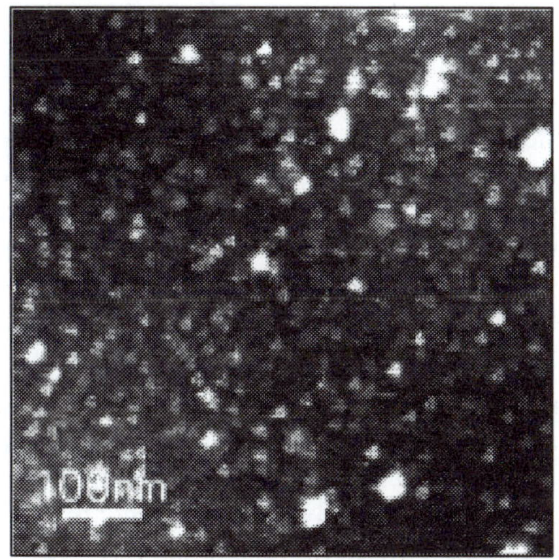

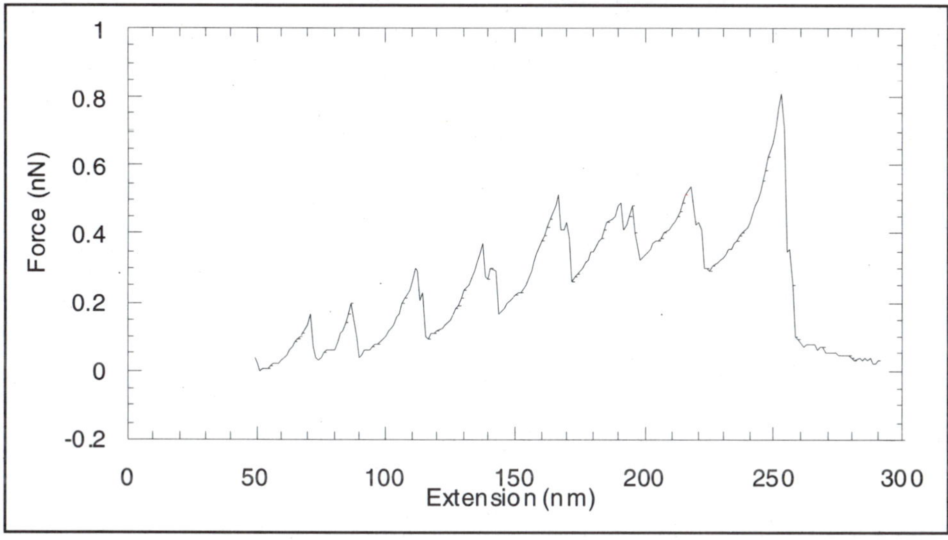

Figure 9.5: A Saw Tooth Force Distance Curve Results when a Titin Protein Molecule is Extended Between the AFM Cantilever Tip and the Mounting Surface. Each minimum of cantilever deflection corresponds to a force maximum where a individual titin domain is ruptured. The distances between minimum peaks correspond to the length of a domain.

periodicity of 25 to 28 nm. The expected distance to unravel a single titin Ig domain is 31 nm (Rief *et al.*, 1997). In addition to proteins, AFM force spectroscopy experiments have been preformed on polysaccharides (Li *et al.*, 1999), DNA (Rief *et al.*, 1999) and alcohols (Li *et al.*, 2000).

Measuring rupture forces within and between molecules by simply retracting the cantilever is possible with most commercial AFM's today. However, in order to extend the capabilities of this technology, a number of modifications to the AFM would be helpful. A low coherence laser to reduce background noise in the force distance curve would be desirable along with a closed loop feedback in the Z piezoelectric ceramic in order to have extremely accurate positioning of the tip in the height domain. Scripting capability where the operator can program movement of the cantilever tip would be another desirable feature. All of these options are currently available in the Agilent 5500 AFM instrument. Additionally user access to programmed applications for either applying a predetermined constant force or force at a constant rate should be integrated into AFM instruments.

Molecular Recognition Force Microscopy (MRFM) is another area where AFM will play a significant role in future biological research. This technology combines topographic imaging with force recognition. By attaching a probe molecule to the cantilever tip, through an 8-30 nm polyethylene glycol tether [Haselgruber, 1995], images of a surface can be obtained simultaneously with recognition of force interactions where the probe molecule interacts with its specific target on the surface. This was first demonstrated by Hinterdorfer's group where an antibody to lysozyme was tethered to the cantilever tip and topographic images, taken in liquid environment, of force recognition was obtained as the tip was scanned over a mica surface of immobilized lysozyme [Raab, 1999]. What actually happens is that the image becomes distorted due to the interaction of the antibody tethered on the tip with the lysozyme coated surface. In Figure 9.6, a topographic image of an avidin coated surface is shown along with a force distance curve taken with a biotin molecule attached by a 30 nm tether to the cantilever tip. In the force curve the initial peak is caused by adhesion as the cantilever tip touches the surface and is retracted. Normally when imaging in air where the humidity is low and when imaging in liquid this adhesion peak is virtually absent. The second peak that is roughly 30 nm from the adhesion peak is due to breaking the bond of biotin tethered to the cantilever interacting with avidin on the surface.

So in principle during sample scanning if the point where the image information is being taken could be separated from the interaction peak we could simultaneously acquire both a topographic and recognition image of a surface. This has been accomplished by the Hinterdorfer group and is shown in Figure 9.7. Both topographic and recognition images of avidin molecules on a mica surface are being obtained simultaneously as biotin tethered to the cantilever tip is scanned over the surface. Recently this system has become available as an option by Agilent Technologies to be added onto their 5500 scanning probe microscope system.

Summary

AFM technology has matured to a point where it can assume an important role in biological research in the twenty first century. The field is in a state of rapid growth, and manufacturers of commercial instruments are eager to incorporate promising new ideas from researchers into commercial products. However, speculation about future development of instrumentation is risky.

Distance [nm]

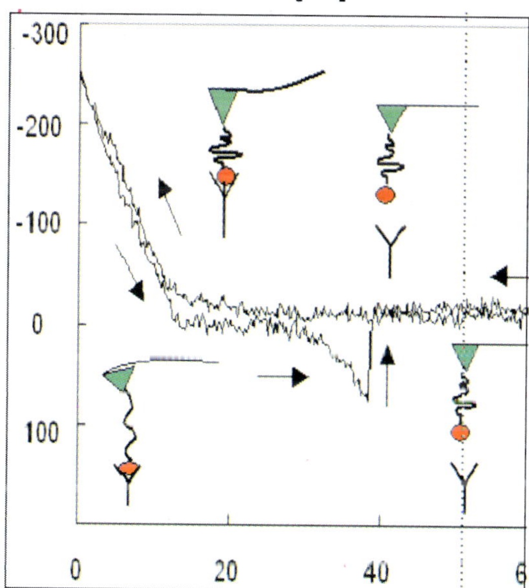

Figure 9.6: A MAC Mode Image of Avidin Protein on a Mica Surface (Left). Image size 450 nm. By contacting the surface with a biotin molecule, tethered to the cantilever tip, the force distance curve (Right) shows the peak due to breaking the biotin-avidin interaction as the tip is retracted from the surface.

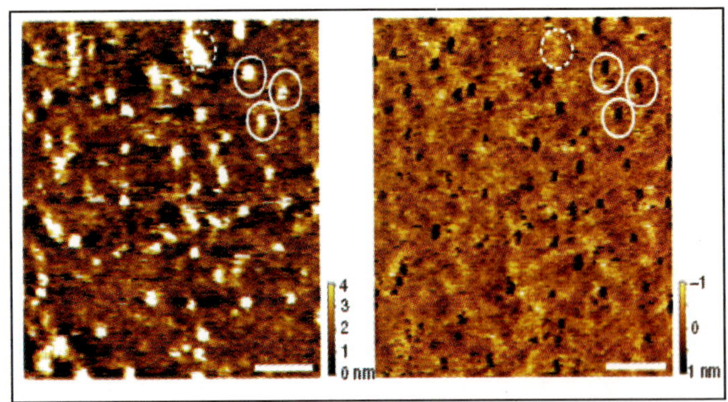

Figure 9.7: Topographic and Recognition Images obtained simultaneously using the PicoTREC System Installed on an Agilent 5500 Scanning Probe Microscope. A mica surface decorated with avidin molecules was scanned with biotin tethered to the cantilever tip. In the topographic image biotin molecules are imaged as white spots (Left). In the recognition image, in the same position, biotin molecules are observed as dark holes (Right). Scan size 500 nm.

Perhaps a better approach is to look at the future of biological research for directions of instrument development. The end of the twentieth century saw the human genome sequenced. In this century, biological research will be devoted to trying to figure out how the protein product of genes interacts to maintain a viable cell and consequently a living organism. As has been pointed out in this paper, this will require methods for identifying interactions between a variety of components and for mapping these features within the living cell. We will all have to wait to see how AFM instrumentation responds to these challenges.

References

Allison, D. P., Kerper, P. S., Doktycz, M. J., Spain, J. A., Modrich, P., Larimer, F. W., Thundat, T., Warmack, R.J., 1996. "Direct Atomic-Force Microscope Imaging of EcoRI Endonuclease", Proc. Natl. Acad. Sci., 93(17), pp.8826.

Allison, D. P., Kerper, P. S., Doktycz, M. J., Thundat, T., Modrich, P., Larimer, F. W., Johnson, D. K., Hoyt, P. R., Mucenski, M. L., Warmack, R.J., 1997."Mapping Individual Cosmid DNAs by Direct AFM Imaging", Genomics, 41, pp.379.

Benoit, M., Gabriel, D., Gerisch, G., Gaub, H. E., 2000. 'Discrete Interactions in Cell Adhesion Measured by Single-molecule Force Spectroscopy," Nat. Cell Biol. 2, pp.3134.

Binnig, G., Rohrer, H., 1982. "Scanning Tunneling Microscopy", Helv. Phys. Acta, 55(6), pp.726.

Binnig, G., Quate, C. F., Gerber, Ch., 1986. "Atomic Force Microscopy," Phys. Rev. Lett., 56, pp.930.

Boland, T., Ratner, B.D., 1995."Direct Measurement of Hydrogen Bonding in DNA Nucleotide Bases by Atomic Force Microscopy", Proc. Natl. Acad. Sci., 92, pp.5297.

Bonnell, D., 2000. "Scanning Probe Microscopy and Spectroscopy: Theory, Techniques, and Applications", 2nd Edition, John Wiley and Sons Inc. New York.

Bustamante, C., Vesenka, J., Tang, C. L., Rees, W., Guthold, M., Keller, R., 1992. "Circular DNA Molecules Imaged in Air by Scanning Force Microscopy," Biochem, 31, pp.22.

Bustamante, C., Rivetti, C., 1996. "Visualizing Protein-nucleic Acid Interactions on a Large Scale with the Scanning Force Microscope," Annu. Rev. Bioph. Biom., 25, pp.395.

Doktycz, M. J., Sullivan, C. J., Hoyt, P. R., Pelletier, D. A., Wu, S., Allison, D. P., 2003. "AFM imaging of bacteria in liquid media immobilized on gelatin coated mica surfaces" Ultramicroscopy, 97(1-4), in press.

Florin, E. L., Moy, V. T., Gaub, H. E., 1994."Adhesion Forces Between Individual Ligand-Receptor Pairs", Science, 264, pp.415.

Hallett, P., Offer, G., Miles, M. J., 1995. "Atomic Force Microscopy of the Myosin Molecule," Biophys. J., 68, pp.1604.

Haselgrubler, T., Amerstorfer, A., Schindler, H., Gruber, H., 1995."Synthesis and Applications of a New Poly (ethylene Glycol) Derivative for the Cross-Linking of Amines with Thiols", Bioconjugate Chemistry, 6, pp.242.

Han, T., Williams, J. M., Beebe, T. P., 1995. "Chemical-bonds Studied with Functionalized Atomic Force Microscopy," Anal. Chim. Acta., 307, pp.365.

Han, W., Lindsay, S. M., Jing, T., 1996. "A Magnetically Driven Oscillating Probe Microscope for Operation in Liquids," Appl. Phys. Lett., 69, pp.4111.

Hansma, H. G., Vesenka, J., Siegerist, C., Kelderman, G., Morrett, H., Sinsheimer, R. L., Elings, V., Bustamante, C., Hansma, P. K., 1992. Reproducible Imaging and Dissection of Plasmid DNA Under Liquid with Atomic Force Microscope," Science, 256, pp.1180.

Hansma, H. G., Laney, D. E., 1996. "DNA Binding to Mica Correlates with Cationic Radius: Assay by Atomic Force Microscopy," Biophys. J., 70, pp.1933.

Hansma, H. G., Kim, K. J., Laney, D. E., Garcia, R. A., Argaman, M., Allen, M. J., Parsons, S. M., 1997. "Properties of Biomolecules Measured from Atomic Force Microscope Images: A Review," J. Struct. Biol., 119 (2), pp.99.

Hansma, P. K., Cleveland, J. P., Radmacher, M., 1994. "Tapping Mode Atomic Force Microscopy in Liquids," Appl. Phys. Lett., 64, pp.1738.

Henderson, E., Haydon, P. G., Sakaguchi, D. S., 1992. "Actin Filament Dynamics in Living Glial Cells Imaged by Atomic Force Microscopy," Science, 257, pp.1944.

Hochachka, P. W., 1999. "The metabolic implications of intracellular circulation," Proc. Natl. Acad. Sci., 96 (22) pp. 12233.

Hoh, J. H., Cleveland, J. P., Prater, C. B., Revel, J-P., Hansma, P. K., 1992."Quantized Adhesion Detected with the Atomic Force Microscope", J. Am. Chem. Soc., 114, pp.4917.

Kasas, S., Thompson, N. H., Smith B. L., Hansma, H. G., Zhu, X., Guthold, M., Bustamante, C., Kool, E. T., Kashlev, M., Hansma, P. K., 1997. "Escherichia coli RNA Polymerase Activity Observed Using Atomic Force Microscopy, Biochemistry, 36, pp.461.

Lantz, M., O'Shea, S. J., Welland, M. E., 1994. "Force Microscopy Imaging in Liquids Using ac Techniques," Appl. Phys. Lett., 65, pp.409.

Lee, G., Arscott, P. G., Bloomfield, V. A., Evans, D. F., 1989. Scanning Tunneling Microscopy of Nucleic Acids," Science, 244, pp.475.

Lee, G. U., Chrisey, L. A., Colton, R. J., 1994a. "Direct Measurement of the Forces Between Complementary Strands of DNA", Science, 266, pp.771.

Lee, G. U., Kidwell, D. A., Colton, R. J., 1994b. "Sensing Discrete Streptavidin Biotin Interactions with Atomic Force Microscopy," Langmuir, 10, pp.354.

Li, H. B., Rief, M., Osterhelt,F., Gaub, H. E., Zhang, X., Shen, J. C., 1999. "Single Molecule Force Spectroscopy on Polysaccharides by AFM Nanomechanical Fingerprint of α-(1-4)-Linked Polysacharides," Chem. Phys. Lett., 305, pp.197.

Li, H. B., Zhang, W., Xu, W., Zhang, H., 2000. "Hydrogen Bonding Governs the Elastic Properties of Poly (vinyl alcohol) in Water: Single-molecule Force Spectroscopic Studies of PVA by AFM," Macromolecules, 33, pp.465.

Lindsay, S. M., Lyubchenko, Y. L., Gall, A. A., Shlyakhtenko, L. S., Harrington, R. E., 1992. "Imaging DNA Molecules Chemically Bound to Mica Surfaces," SPIE Proceedings 1639, pp.84.

Ludwig, M., Moy, V. T., Rief, M., Florin, E. L., Gaub, H. E., 1994."Characterization of the Adhesion Force Between Avidin- Functionalized Afm Tips and Biotinylated Agarose Beads", Microscopy Microanalysis Microstructures, 5, pp.321.

Lutwyche, M.I., Despont, M., Drechsler, U., Durig, U., Haberle, W., Rothuizen, H., Stutz, R., Widmer, R., Binnig, G., Vettiger, P., 2000. "Highly Parallel Data Storage System Based on Scanning Probe Arrays", Appl. Phys. Lett., 77 (20), pp.3299.

Lyubchenko, Y. L., Gall, A. A., Shlyakhtenko, L. S., Harrington, R. E., Oden, P. I., Jacobs, B. L., Lindsay, S. M., 1992. "Atomic Force Microscopy Imaging of Double Stranded DNA and RNA," J. Biomolecular Structural Dynamics, 9, pp.589.

Marszalek,P. E., Lu, H., Li, H. B., Carrion-Vazquez, M., Oberhauser, A. F., Schulten, K., Fernandez, J. M., 1999. "Mechanical Unfolding intermediates in Titin Molecules," Nature, 402,pp. 100.

Morris, V. J., Kirby, A. R., Gunning, A. P., 1999. "Atomic Force Microscopy for Biologists", Imperial College Press.

Niemeyer, C.M., 2001. "Nanoparticles, Proteins, and Nucleic Acids: Biotechnology Meets Materials Science," Agnew. Chem. Int. Ed., 40, pp.4128.

Oberhauser, A. F., Marszalek, P. E., Erickson, H. P., Fernandez, J. M., 1998. "The Molecular Elasticity of the Extracellular Matrix Protein Tenascin," Nature 393, pp.181.

Quist, A. P., Bjorck, L. P., Reimann, C. T., Oscarsson, S. O., Sundquist, B. U. R., 1995. "A Scanning Force Microscopy Study of Human Serum Albumin and Porcine Pancreas Trypsin Adsorption on Mica Surfaces," Surface Sci., 325, pp.406.

Raab, A., Han, W., Badt, D., Smith-Gill, S. J., Lindsay, S. M., Schindler, H., Hinterdorfer, P., 1999. "Antibody Recognition Imaging by Force Microscopy", Nat. Biotechnol., 17, pp.902.

Radmacher, M., Tillmann, R. W., Fritz, M., Gaub, H. E., 1992. "From Molecules to Cells: Imaging Soft Samples with Atomic Force Microscope," Science 257, pp.1900.

Radmacher, M., Fritz, M., Cleveland, J. P., Walters, D. R., Hansma, P. K., 1994."Imaging Adhesion Forces and Elasticity of Lysozume Adsorbed on Mica with the Atomic Force Microscope," Langmuir, 10, pp.3809.

Rief, M., Gautel, M., Oesterhelt, F., Fernandez, J. M., Gaub, H. E., 1997. "Reversible Unfolding of Individual Titin Immunoglobulin Domains by AFM", Science, 276, pp.1109.

Rief,M., Clausen-Schaumann, H., Gaub, H. E., 1999. "Sequence-dependent Mechanics of Single DNA Molecules," Nat. Struct. Biol., 6, pp.346.

Rotsch,C., Radmacher, M., 2000. "Drug-induced Changes of Cytoskeletal Structure and Mechanics in Fibroblasts: An Atomic Force Microscopy Study," Biophys. J. 78(1), pp.520.

Sonnenfeld, R., Hansma, P. K. 1986. "Atomic-resolution Microscopy in Water," Science,232, pp.211.

Thompson,N. H., Kasas, S., Smith, B., Hansma, H. G., Hansma, P. K., 1996. "Reversible Binding of DNA to Mica for AFM Imaging," Langmuir, 12, pp.5905.

Thundat, T. Allison, D. P., Warmack, R. J., Ferrell, T. L., 1992. "Imaging Isolated Strands of DNA Molecules by Atomic Force Microscopy," Ultramicroscopy, 42-44, pp.1101.

Thundat, T., Allison, D. P., Warmack, R. J., Brown, G. M., Jacobson, K. B., Schrick, J. J., Ferrell, T. L., 1992. "Atomic Force Microscopy of DNA on Mica and Chemically Modified Mica," Scanning Microsc. 6, pp.911.

Tskhovrebova, L., Han, W., Trinick, J., 2000. "Studies of Mechanical Unfolding of Single Titin Molecules Using AFM AC Mode," Biophys. J., 78, pp.2627.

Wang, H. D., Bash, R., Yodh, J. G., Hager, G. L., Lohr, C., Lindsay, S. M., 2002. "Glutaraldehyde Modified Mica: A New Surface for Atomic Force Microscopy of Chromatin," Biophys. J. 83 (6), pp.3619.

Watanabe, Y., Sakurai, T., 2000. "Advances in Scanning Probe Microscopy", Springer-Verlag, New York..

Whitesides, G. M., Grzybowski, B., 2002. "Self-Assembly at All Scales," Science 295, pp.2418.

Xia, Y., Rogers, J. A., Paul, K. E., Whitesides, G. M., 1999. "Unconventional Methods for Fabricating and Patterning Nanostructures," Chem. Rev. 99(7), pp.1823.

Chapter 10

Neural Network Based Optical Pattern Recognition and Generation

Arbab Ali Khan[1] and Ahmed Shuja[2]
Department of Electronic Engineering,
International Islamic University, Islamabad, Pakistan
E-mail: [1]arbab.ali@iiu.edu.pk, [2]ahmed_shuja@iiu.edu.pk

ABSTRACT

A new neural network based optical pattern recognition and generation scheme is presented. It recognizes binary patterns similar to the schemes presented so far, with an additional power to generate a pattern related to the input massively linking two data basis of stored and related patterns. Both input and related output patterns are stored in same associative memory. This scheme utilizes Hopfield like neural network model and the experimental setup uses holographic gratings. Light while passing through the holographic memories performs the required calculations and is detected by cameras. Their outputs signals are passed through an electronic comparator that generates the final output.

Keywords: *Micro-electronics, Optical pattern, Neural network, Holographic gratings, Electronic comparator.*

Introduction

The need for ultra fast pattern recognition systems is self-evident and thus the field has been very active during the last couple of decades. Pattern recognition, considered analogous to human brain, is content-addressable memory system. An important aspect of human brain is that it can associate information databases about a pattern after identification that lays the basis for thought mechanism which is information processing amongst associated databases. The schemes presented so far recognize the given pattern, may be from some partial information. There are, however, a few those which generate a related pattern. Related pattern may provide the name of the identified person or object or it can be some alphanumeric string that is linked to the information file of the recognized pattern or simply some other pattern.

Several algorithms for the purpose of pattern recognition utilizing neural networks (Hopfield *et al.*, 1982; and Kohonon *et al.*, 1984) have been proposed so far. Optical implementation schemes of some of those algorithms due to inherent nature of massive parallelism can establish multiple interconnections and thus can handle large amount of data and data processing in an associative manner. The optical models, therefore, have proved to be highly promising and efficient in solving these types of fuzzy problems than their electronic counter part.(Yeh *et al.*, 2004; Rizvi *et al.*, 1994; Wang *et al.*, 1991; Yu *et al.*, 1990; Yu *et al.*, 1991; Song *et al.*, 1990; Gao *et al.*, 1997; Psaltis *et al.*, 1985; Awwal *et al.*, 2002;) Many optical implementations use grating structures (Yeh *et al.*, 2004; Rizvi *et al.*, 1994; Wang *et al.*, 1991; Khan 1998; Khan 1994; White 1988). In this paper we propose a neural network based algorithm and its optical implementation using grating structures that recognizes a given input image and generates a pattern related to it stored in the same associative memory.

A New Neural Network

An algorithm presented by Rizvi and Zubairy (Rizvi *et al.*, 1994) for optical pattern recognition using associative memory with novel thresholding technique has recently been implemented experimentally (Deng *et al.*, 2005). Same type of experimental setup can be utilized with a little modification for the implementation of this algorithm. However, in this paper, we incorporate not only the *on-state* but also the *off-state* pixels in the recognition process that makes the scheme more powerful and also eliminates the need of thresholding the output matrix. Though utilization of *on-* and *off-state* pixels in the process requires twice the memory for its implementation, however, it makes the scheme more robust to partial or even to some extent to wrong information at the input.

The subtractions needed, can be performed electronically. Here the partial input information corresponds to the missing values at both the inputs to the respective memory matrices for *on-* and *off-state* pixels and contributes nothing in the reconstruction process. In the optical implementation, partial input information means portions of the input that are opaque for both *on-* and *off-state* memories. Wrong information, that adversely affects the recognition process, means the *on-state* pixels in the input that were *off* in the stored pattern and vice versa. For the faithful recognition we, however, impose a condition on the choice of the stored patterns that the number of *on-* and *off-state* pixels at a particular position in the stored patterns remains roughly equal. An input generates at the output a related-pattern of that stored-pattern for which the input has minimum hamming distance. This scheme can be extended very easily for multicolored pattern recognition using the technique presented by Arbab and Zubairy (Khan *et al.*, 1994). Similar to that, a multicolored version can easily retrieve the desire result among a low hamming distance database and is more immune to errors and partial inputs. There are three categories of patterns in this scheme, first:

acting as stored-patterns (S) from which any one is to be recognized; second: their related-patterns (R) that the system has to produce as the output, both of these are stored in the same memory and the third: given as input-pattern (I) to the system for recognition. Here S, R and I are two dimensional arrays of binary bits representing patterns and can be one dimensional for signal processing.

As an illustration, we consider stored patterns as two dimensional 4×4 matrices with their elements (pixels) acting as neurons, as each pixel effects every other pixel in the system. The size of the stored and the related patterns can be different and there is no restriction on size of the patterns. Assuming that there are P pairs of images stored in the memory and stored-patterns are represented by one-dimensional M pixels arrays and their related-patterns by N pixels arrays, such that S_m, $1 \leq m \leq M$, and R_n, $1 \leq n \leq N$, equals either 0 or 1 corresponding to an *off-* or *on-state* of the pixel respectively. The size of the memory matrix is $(M \times N)$ and its expression becomes:

$$V_{ij} = \sum_{p=1}^{P} S_i^p R_j^p, \tag{1}$$

where,

Vij is the value of the pixel at i^{th} row and j^{th} column of the memory matrix V, Sip is the ith pixel of stored input-pattern and Rjp is j^{th} pixel of the related output-pattern of pth pair respectively.

Second memory matrix for the off-state pixels (negative image) is formed by the same expression reversing 0s and 1s in the patterns. It is also important to note that the memory matrices can easily be upgraded for new pair of patterns by just adding new calculated weights to the respective elements of the previously prepared memory matrix and there is no need to recalculate the whole memory matrix. In the optical implementation, however, the memory needs to be refreshed with new values.

The reconstruction process is such that an input pixel in the *on-state* generates all the *on-state* pixels of those related output-patterns for which that particular pixel of the stored input-patterns was in the *on-state*. Likewise an *off-state* pixel in the input generates all the *off-state* pixels of those output-related patterns for which their associated input-stored patterns have the same pixel in *off-state*. Thus a related image in the memory is recalled as many times as the number of *on-state* pixels common with the input and the stored image, for *on-state* memory. These recalled images summed together generate multilevel (gray-scale) image. The image form the memory for *off-state* pixels is subtracted from the image that is contributed by memory for *on-state* pixels to generate the output by assigning *1*s to the positive values and *0*s to the negative values. The relationship for the jth pixel of the gray-scale output array ${}^{g}O$ with the input and the memory is:

$${}^{g}O_j = \sum_{i=1}^{M} I_i V_{ij}, \tag{2}$$

where,

I_i is the ith pixel of the input given to the system. This way every *on-state* pixel I_i $1 \le i \le M$, of an input I, recalls every related output image for which the stored image has the same *on-state* pixel at that position. The overall recall relation can be written as

$$^{g}O_{ij} = \sum_{p=1}^{P} \sum_{k=1}^{K} \sum_{l=1}^{L} I_{kl} S_{kl}^{p} R_{ij}^{p}, \tag{3}$$

where,

K and L are the maximum number of rows and columns in the stored patterns respectively.

The same is true for recognition by off-state pixels and their related memory that generates a second similar image.

The final state of an output pixel at a particular position is obtained by comparing the respective pixels of the two output matrices. An output pixel is assigned on-state if gO jon>gOjoff and off-state if gO jon<gOjoff. In case gO jon=gOjoff, the output pixel can take the same state as in the input given to the system for recognition contributing in the successive iterative process. This comparison can be done by an electronic subtractor. Thus the system going through the recognition process retrieves the complete output pattern.

The recognition process requires a vector-matrix multiplication of a (1×M) vector and a (M×N) matrix and is done optically; however, comparison of two (1×N) arrays can easily be done electronically. To illustrate the model, we consider stored patterns as two dimensional 4×4 matrices

$$\begin{bmatrix} 1 & 1 & 1 & 1 \\ 0 & 0 & 0 & 0 \\ 1 & 1 & 1 & 1 \\ 0 & 0 & 0 & 0 \end{bmatrix}, \begin{bmatrix} 1 & 0 & 1 & 0 \\ 1 & 0 & 1 & 0 \\ 1 & 0 & 1 & 0 \\ 1 & 0 & 1 & 0 \end{bmatrix}, \begin{bmatrix} 0 & 1 & 0 & 1 \\ 0 & 1 & 0 & 1 \\ 0 & 1 & 0 & 1 \\ 0 & 1 & 0 & 1 \end{bmatrix}, \begin{bmatrix} 0 & 0 & 0 & 0 \\ 1 & 1 & 1 & 1 \\ 0 & 0 & 0 & 0 \\ 1 & 1 & 1 & 1 \end{bmatrix},$$

and the related-output patterns as 4×3 matrices:

$$\begin{bmatrix} 0 & 0 & 0 & 1 \\ 0 & 0 & 1 & 1 \\ 0 & 1 & 0 & 1 \end{bmatrix}, \begin{bmatrix} 0 & 0 & 1 & 0 \\ 1 & 0 & 0 & 1 \\ 1 & 1 & 1 & 1 \end{bmatrix}, \begin{bmatrix} 1 & 1 & 0 & 1 \\ 0 & 1 & 1 & 0 \\ 0 & 0 & 0 & 0 \end{bmatrix}, \begin{bmatrix} 1 & 1 & 1 & 0 \\ 1 & 1 & 0 & 0 \\ 1 & 0 & 1 & 0 \end{bmatrix}$$

The memory is thus *16×12* matrix. and we give the following as input which has minimum hamming distance of three from third stored pattern where as it is five, eleven and nine from first, second and fourth input-stored patterns. Further from the third stored pattern pixels *1*st, *3*rd and *12*th are wrong information, where as *8*th is missing from *off-state* input and *11*th is missing from *on-state* input. The contributions from the *on-state* memory and *off-state* memory and the result of comparison to get the final binary output are as follows, thus the algorithm retrieves, as desired, the third related-output pattern:

$$\underset{\text{Input}}{\begin{bmatrix} \mathbf{1} & 1 & \mathbf{1} & 1 \\ 0 & 1 & 0 & 1 \\ 0 & 1 & 0 & \mathbf{0} \\ 0 & 1 & 0 & 1 \end{bmatrix}} \Longrightarrow \begin{bmatrix} 6 & 6 & 4 & 8 \\ 4 & 6 & 8 & 6 \\ 4 & 6 & 4 & 6 \end{bmatrix} - \begin{bmatrix} 5 & 5 & 9 & 3 \\ 9 & 5 & 3 & 7 \\ 9 & 7 & 9 & 7 \end{bmatrix} \Longrightarrow \underset{\text{Output.}}{\begin{bmatrix} 1 & 1 & 0 & 1 \\ 0 & 1 & 1 & 0 \\ 0 & 0 & 0 & 0 \end{bmatrix}}$$

Each element of the memory mask is of the same size as that of input pixel and has a hologram that optically connects the pixels in a massively parallel manner. Thus light passing through a transparent pixel of the input, illuminates one hologram of the memory mask behind it, which generates a pattern at the output. The intensity of a pixel at the output plane depends on the number of stored patterns in which a pixel at the position illuminating the hologram was in *on-state*. All illuminated holograms superimpose their contributions to form a gray level matrix at the output plane. A similar contribution comes from memory for *off-state* pixels at another output plane. Both the contributions are collected by cameras whose outputs are compared electronically to generate the final output.

The schematic diagram is shown in Figure 10.1. There are two inputs one for *on-state* and the other (which is just the inverse) for *off-state* pixels. The holographic associative memories for *on-* and *off-state* pixels produce the first orders and their desired intensities at the output planes. Collimated light sources illuminate the input image transparencies in front of the memory masks. The opaque pixels of the input do not let the adjacent memory pixels to contribute anything to the output.

Illuminated memory pixels diffract light in the first order according to the weights of the pixels at the output plane. The convex lenses take the Fourier transform (J.W. Goodman, 1998) to produce the patterns at a reasonable distance. The images are captured by CCD-cameras and are compared using a computer to display the final result.

A general expression characterizing the transmittance $t_{kl}(x,y)$ of a pixel (k,l) of dimensions $2a \times 2a$ of the memory filter in Cartesian coordinates can be obtained by suitably generalizing the results of diffraction theory in reference (Rizvi *et al.*, 1994) which give:

$$t_{kl}(x,y) = \frac{1}{W} \sum_{p=1}^{P} \sum_{i=1}^{I} \sum_{j=1}^{J} S_{kl}^{p} R_{ij}^{p} \times \left[1 + C \cos \frac{2\pi}{\lambda D} \{2a(i-l)x + (E - 2a(j+J-k))y\} \right], \tag{4}$$

where,

S_{kl}^{P} and R_{ij}^{P} are the states of pixels of kth row and l^{th} column and i^{th} row and j^{th}

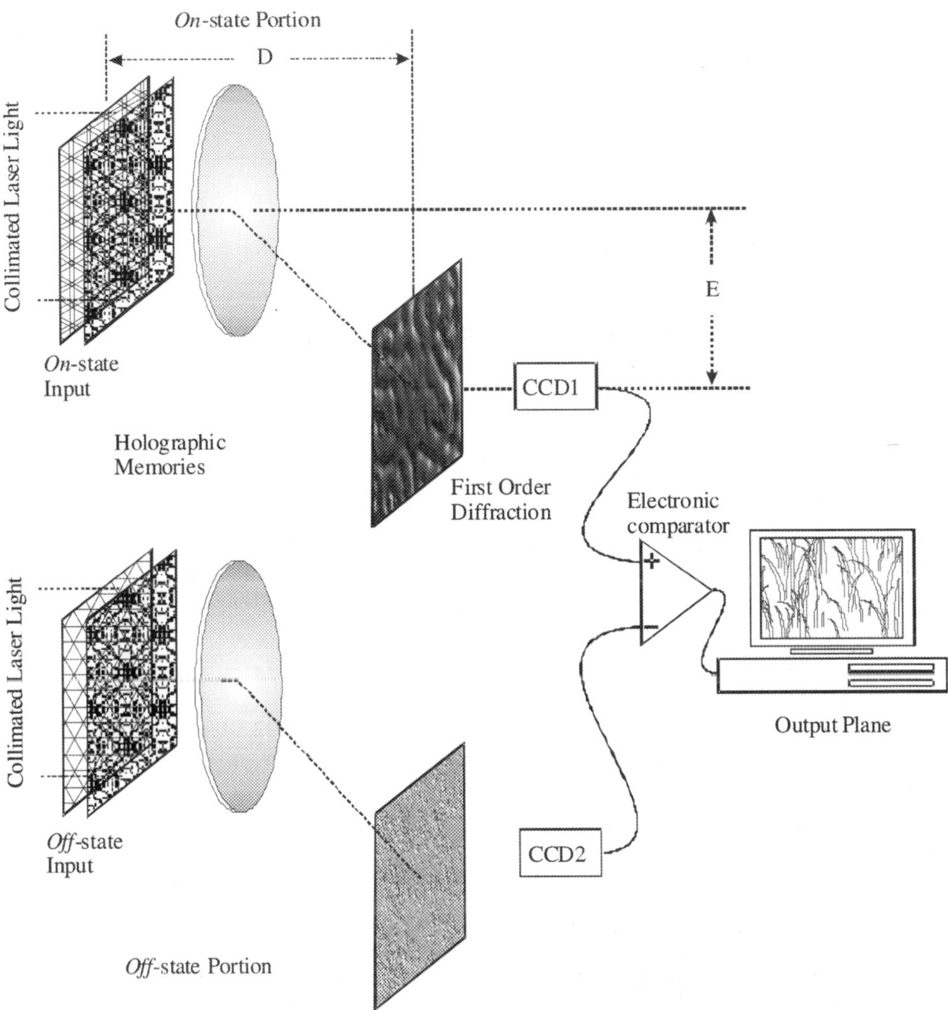

Figure 10.1: Collimated Light Source from the Left Illuminates the Input Transparencies, Adjacent are the Memory Masks whose Fourier Transform by the Lens Reaches the Output Plane, where CCD Cameras Capture the Images which are Compared Electronically and Displayed by a Computer

column of the stored and related patterns of pth pair, P is the total number of pairs stored in the memory, I and J are the maximum number of rows and columns in the related patterns respectively such that $I{\times}J{=}N$, D is the distance between the memory and the output plane (horizontal distance) and E is the distance between 0th and the 1st orders, *i.e.*, vertical distance to protect overlapping of diffraction orders. The constant C is the contrast of the fringes whose value depends upon fringe visibility and W is the normalization factor which equals to the largest value in the memory matrix. The transmission function of the complete memory with center of the transparency as the origin comes out to be (Khan *et al.*, 1994):

$$t(x,y) = \frac{1}{W}\sum_{p=1}^{P}\sum_{i=1}^{I}\sum_{j=1}^{J}\sum_{k=1}^{K}\sum_{l=1}^{L} S_{kl}^{p} R_{ij}^{p} \times \quad (5)$$
$$rect\left[\frac{x+(J+1-2l)a}{2a}, \frac{y+E+(2k-J-1)a}{2a}\right] \times$$
$$\left(1 + C\cos\frac{2\pi}{\lambda D}\left[(J+1-2l)ax + \{E+(2k-J-1)a\}y\right]\right)$$

Light intensity at the output plane can be calculated by the application of the Fraunhoffer's diffraction formula. (J.W. Goodman, 1998) It is worth mentioning that the optical setup given in reference (Khan *et al.*, 1998) section 3, for vector matrix multiplication can also be adopted for the implementation of this algorithm.

In this paper we have given an algorithm for related pattern generation and have shown that an associative memory is capable of generating related output patterns, not only electronically but also optically. The optical implementation of the memory for the reconstruction of related pattern is given rigorous treatment. Holographic techniques or spatial light modulators can be utilized for memory fabrication, the later being much flexible.

References

Awwal, A.A. S., Tang, H., Gudmundsson, K. S., Khan, J., 2002. "Uni-complex and bi-complex representation for associative memory with superior retrieval", Proc. SPIE 4788, pp. 159.

Deng, Z., Qing, D.–K., Hemmer, P. R., Zubairy, M. S., 2005. "Implementation of optical associative memory by a computer-generated hologram with a novel thresholding scheme", Opt. Lett. 30, pp.1944.

Goodman, J.W., 1998. Introduction to Fourier Optics (McGraw-Hill, New York).

Gao, S., Yang, Feng, J. Z., Zhang, Y., 1997. "Implementation of a large-scale optical neural network by use of a coaxial lenslet array for interconnection", Appl. Opt. 36, pp.4779.

Hopfield, J. J., 1982. "Neural networks and physical system with emergent collective computational abilities", Proc. Natl. Acad. Sci. U.S.A. 79, pp.2552.

Khan, A.A., Rizvi, A. A., Zubairy, M. S., 1994. "Colored pattern recognition with a neural network model", Appl. Opt. 33, pp.5467.

Khan, Arbab A., Zubairy, M. S., 1998."Optical computing without using optical elements", Pure and Appl. Opt. 7, pp.1125.

Kohonon, T. 1984. Self-Organization and Associative Memory (Springer-Verlag, New York).

Psaltis, D., Farhat, N., 1985. "Optical information processing based on an associative-memory model of neural nets with thresholding and feedback", Opt. Lett. 10, pp.98.

Rizvi, A.A., Zubairy, M. S., 1994. "Implementation of associative memory using grating structures", Appl. Opt. 33, pp.3642.

Song, S. H., Park, S.–C., Lee, S. S., 1990. "Optical implementation of a quadratic associative memory by using the polarization-encoding process", Opt. Lett. 15, pp.1389.

White, H. J., Wright, W. A., 1988. "Holographic implementation of a Hopfield model with discrete weightings", Appl. Opt. 27, pp.331.

Wang, X.–M., Mu, G.–G., Zhang, Y.–X., 1991. "Optical associative memory using an orthogonalized hologram", Opt. Lett. 16, pp.100.

Yu, F. T. S., Lu, T., Yang, X., Gregory, D. A., 1990. "Optical neural network with pocket-sized liquid-crystal televisions", Opt. Lett. 15, pp.863.

Yu, F. T. S., Yang, X., Lu, T., 1991. "Space–time-sharing optical neural network", Opt. Lett. 16, pp.247.

Yeh, S. L., Lo, R. C., Shi, C.Y, 2004. "Optical implementation of the Hopfield neural network with matrix gratings", Appl. Opt. 43, pp.858.

Chapter 11

Status of Micro and Nanoelectronics Research and Development in Indonesia

Siti Amini Ika[1], *Pepen Arifin*[2] *and Hiskia Sirait*[3]

[1]*Center for Nuclear Fuel Technology-National Nuclear Energy Agency, Tangerang, Indonesia*
E-mail: amini@batan.go.id, minika_s@yahoo.com

[2]*Department of Physics, Institute Technology of Bandung, Bandung, Indonesia.*
E-mail: pepen@fi.itb.ac.id

[3]*Research Center for Electronics and Telecommunication, Indonesian Institute of Sciences, Bandung, Indonesia.*
E-mail: hiskia@ppet.lipi.go.id, hiskia_sirait@yahoo.com

ABSTRACT

Status of recent research and development efforts on micro-nano sciences and technology in Indonesia is presented. Research and development in this field generally has strong relationship with partners abroad in terms of collaborative research. Fabrications of silicon quantum dots were done by several researchers by different methods. Since many areas of chemistry have important roles in the material process, the material chemistry has high potential for application in the micro- and nano-electronics devices especially for the Micro Electronic Mechanical System (MEMS) and Nano Electronic Mechanical System (NEMS). Also, many researches in the field of catalyst with wide range of applications have been conducted attractively such as for developing fuel cell along with development of their membranes etc. Many potential materials such as silicon, zeolites, clays, minerals have been targeted to be explored with various ways, and thus their characterizations need precise instruments that are based on the electronical-system with micro or nano devices to perform accurate measurements. Many activities conducted in various institutions are in the field of nanoelectronics, semi-fabrication of nanodevices such gas sensors etc. Unfortunately, as far as the national program is concerned there have been quite a few links with the industrial sectors. The MOST: Ministry of Sciences and Technology and Institutions also plan to support and strengthen the links with the industrial sector in these fields. The activities of the present Workshop are expected to provide opportunity for scientist–to–scientist contact and interaction; familiarizing participants on the latest developments and techniques in the subject areas. The academic-research-industry/business and government interaction under globally governed management in the developing countries is very important. We expect fruitful results through the linked NAM S&T-Industry Network Program.

Keywords: *Quantum Dots, MEMS, Sensors, Nanomaterials; R&D Status in Indonesia.*

Introduction

One of other function of the State Ministry of Research and Technology (in Indonesian namely KNRT) is to coordinate the programs in the field of research, science and technology. There are 7-Non Departmental Research Agencies (LPND) under it, namely:

1. Agency for Assessment and Application of Technology (BPPT)
2. Indonesian Institute of Sciences (LIPI)
3. National Nuclear Energy Agency (BATAN)
4. Nuclear Energy Control Board (BAPETEN)
5. National Institute of Aeronautics and Space (LAPAN)
6. National Coordination Agency for Surveys and Mapping (BAKOSURTANAL)
7. National Standardization Agency of Indonesia (BSN).

EIJKMAN Molecular Biology Institute is another research agency, under the ministry.

In some cases, the performed researches are grouped in form of collaboration work between universities and the research agency's researchers with limited supports from the government, private and/or industrial sectors. Some research centers had been starting their research focus on micro/nano-materials and electronics based on micro/nano-systems. The Ministry for Research and Technology started to initiate a budget support for the nano-sciences and technology research and development programme in Indonesia that is being carried out in the LPND (the Non Departmental Research Institutes/Agencies) and in the universities laboratories since 2004. In the year 2005, the activities were limited to the assessment of nano-science and nano-technology trends, study and preparation for technology roadmap of nano-science and technology R&D, and to explore some practical studies in synthesizing the three subjects *i.e.*: prototype for nano-particles equipments, prototype for micro-devices (MEMS sensor for CO-gas, glucose in blood, Quantum Dots etc.) and prototype for coating materials. The total budget is about US$ 20, 000- excluding the internal institutional budget. Further, it has not much contribution from industries yet. Other studies related with MEMS or NEMS or nano-sciences and technology were also carried out independently within institutional cooperation. This paper will focus mainly on the micro/nanoelectronics research activity in Indonesia, which has been coordinated by the Ministry of Research and Technology till now. It describes the research activities especially in field of MEMS or electronic devices and micro-nano materials.

The main objective is to share knowledge and to extend our capability by broadening networks and possibly collaboration within the countries abroad. Most micro/nano-fabrication techniques have their roots in the standard fabrication methods developed for the semiconductor industry. In most of electronic products, such as integrated circuits (IC's), high performance is required leading to further shrinkage of transistor down to the less than 0.1mm and higher integration. There are alternative devices which enable replacement of today's electronic devices *i.e.*:

- ✰ Solid state quantum effect and single electron devices
- ✰ Molecular electronic devices

Quantum effect and single electron devices in solid state system are being done by most research groups while the latter, molecular electronic devices are relatively new approach that would change both the operating principle and the materials used in electronic devices. More recently, nanometer-sized structures have attracted an enormous amount of interest. This is mainly due to their unique electrical, magnetic, optical, thermal, and mechanical properties. These could lead to a variety of electronic, photonic, and sensing devices with a superior performance compared to their macro counterparts.

Several important *nanofabrication* techniques are currently under intense investigation. Although e-beam and other high resolution lithography methods can be used to fabricate nanometer-sized structures, their serial nature and/or cost preclude their widespread application. This has forced investigators to explore alternatives, and recent innovations in the area of *micro/nanofabrication* have created a unique opportunity for manufacturing structures in the nanometer–millimeter range.

There are several significant nanotechnology research and development efforts underway. The various important MEMS fabrication techniques are commonly used to build various micro-devices (micro-sensors and microactuators) (Kovacs *et al.*, 1998).

Fabrications of silicon quantum dots (QD) were done by several researchers by different methods. A focus will be on processing and manufacturing of new devices and applications. Since nano-structured materials are generally metastable with respect to their larger-structured counterparts, the stability of particle assembly and other nanostructures under the effects of thermal, magnetic, optical and chemical factors and methods to stabilize them also will be surveyed. Major micro-fabrication techniques used most frequently in the manufacturing of micro/nanostructures will be discussed. Some of these techniques such as thick/thin-film deposition and etching are common between the micro/nano and microchip fabrication disciplines.

However, several other studies that are related to the possible materials used in the micro/nanofabrication area will also be described. Recent research in Indonesia related to micro/nanoelectronic fields that have been supported by the Ministry of Research and Technology covers semiconductor QD for laser and optoelectronic applications and micro devices prototype for MEMS *i.e.* CO-gas sensor and glucose sensor. The paper also covers other related research elsewhere in Indonesia and of possible research on nano materials that has promise to be explored in the future.

Designed Activities

Quantum Dots

Many CVD techniques have been used to produce epitaxial growth. The most common for silicon is thermal chemical vapor deposition or vapor phase epitaxy (VPE).

Metallorganic chemical vapor deposition (MOCVD) and molecular beam epitaxy (MBE) are the most common techniques for growing high quality III-V compound layers with nearly atomic abrupt interfaces. The former uses vapors of organic compounds with group III atoms such as trimethylgallium [$Ga(CH_3)_3$] and group V hydrides such as AsH_3 in a CVD chamber with fast gas switching capabilities. The latter typically uses molecular beams from thermally evaporated elemental sources aiming at the substrate in an ultrahigh vacuum chamber. In this case, rapid on/off control of the beams is achieved by using shutters in front of the sources.

QD GaN has been successfully grown using Molecular Beam Epitaxy (MBE) (Shen *et al.*, 1998; Pierre *et al.*, 2001) and Metal Organic Chemical Vapor Deposition (MOCVD) (Morcoc *et al.*, 2001; Tanaka *et al.*, 1996; Ramvall *et al.*, 2000). The quality is expected to be better than that of GaAs dots, but in this case the optimum quality level has not yet been achieved due to relatively low dot-density (5.10^9–5.10^8 cm^{-2}) with the growth temperature, which is relatively high *i.e.* 1050-1100°C with conventional MOCVD method. Considerable film desorption occurs, which causes increase point defects and would decrease dot's distribution (Morcoc *et al.*, 2001). The Qualified QD for instant determined by low defect density ($<10^8$ cm^{-2}), high QD-density ($>10^{11}$ cm^{-2}), also by Dot forms and sizes that would fix to increase its exitonic-binding energy c.a. dot diameter of higher than 4 nm. The mentioned quality is dependent on some determinant growth parameters such as temperatures, surfactant concentration, gas-carrier concentration and time (Tanaka *et al.*, 1996; Shen *et al.*, 1998).

In this research, GaN quantum dot was done by Plasma Assisted MOCVD method by the research group in the Physics of Electronic Materials Laboratory at Institute Technology Bandung (ITB). This method is the result of conventional MOCVD development which is possibly used at low temperatures *i.e.* 600-700° C. It is expected that GaN-QD resulted from this process would show better quality.

MEMS: Sensors

This research is being performed by the research group from Research Centre for Electronics and Telecommunication (PPET-LIPI)-Bandung. Two types of sensors *i.e.* gas sensor and biosensor would be designed. Gas sensor is proposed to determine carbon monoxide gas (CO) and will be developed for others gases application also. Efforts have been made to make miniaturized gas sensor. The micro-machined gas sensors comprise the advantages of micro-size structures in identical, highly uniform and low power consumption.

Biosensor is applied for the determination of glucose concentration in blood or on the other way it can also be applied for analysis of others specific biomaterials. It is as a chemical sensor, which comprises of 3 basics components *i.e.* receptor (bio-component), transducer (physical component) and separator (membrane or some specific coating) (Andreas *et al.*, 2003). Receptor can be made from doped metal oxide or organic polymer compounds which can be interacted with analyte; these may contain enzyme, antigen, antibody, bacteria and nucleic acids. The following figure shows the biosensor principle mechanism.

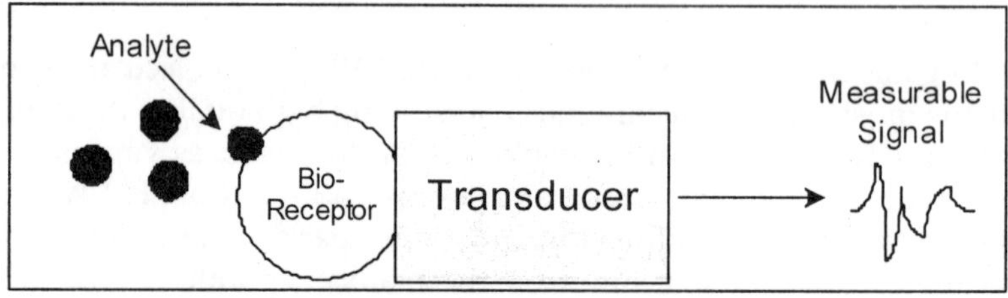

Figure 11.1: Biosensor Principle Mechanism

Particularly, the biosensor is designed for the use of glucose oxidase enzyme (GOD) coated–electrode to detect glucose, as shown in Figure 11.2.

$$\text{Glucose} + O_2 \xrightarrow{\text{Glucose Oxidase}} \text{Gluconic Acid} + H_2O$$

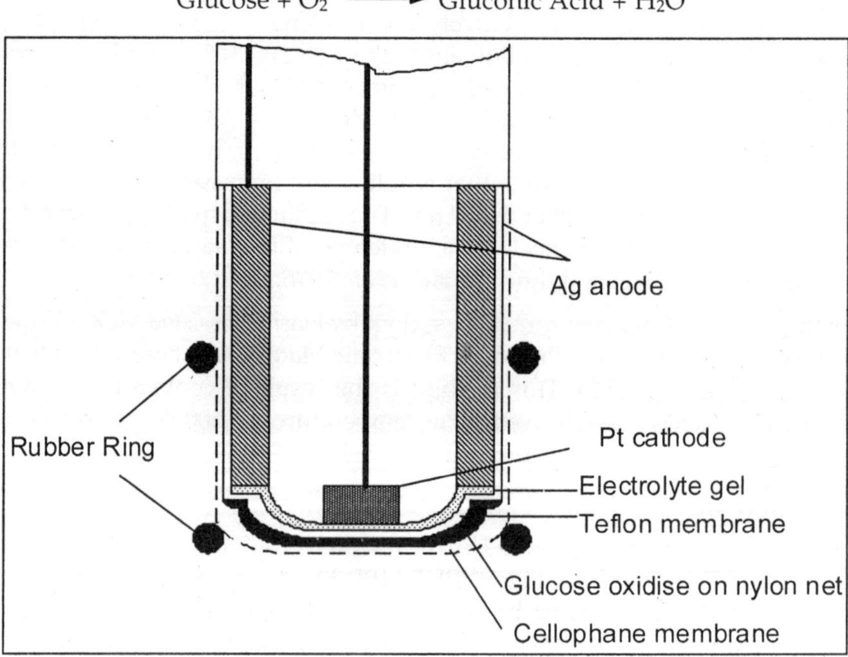

Figure 11.2: Scheme of Glucose Electrode

The work covers micro-fabrication technology *e.g.* fabrication of Integrated Circuit (IC), which is used or compatible with the micro-fabrication process of electrochemical sensor.

Figure 11.3 shows the flow diagram of IC fabrication process. Some sensors (chemical or bio sensors) have been developing recently including ion selective electrodes, gas sensor, glucose sensor and biosensor that can be used for the determination of lactic acid etc. (Depsey *et al.*, 1997).

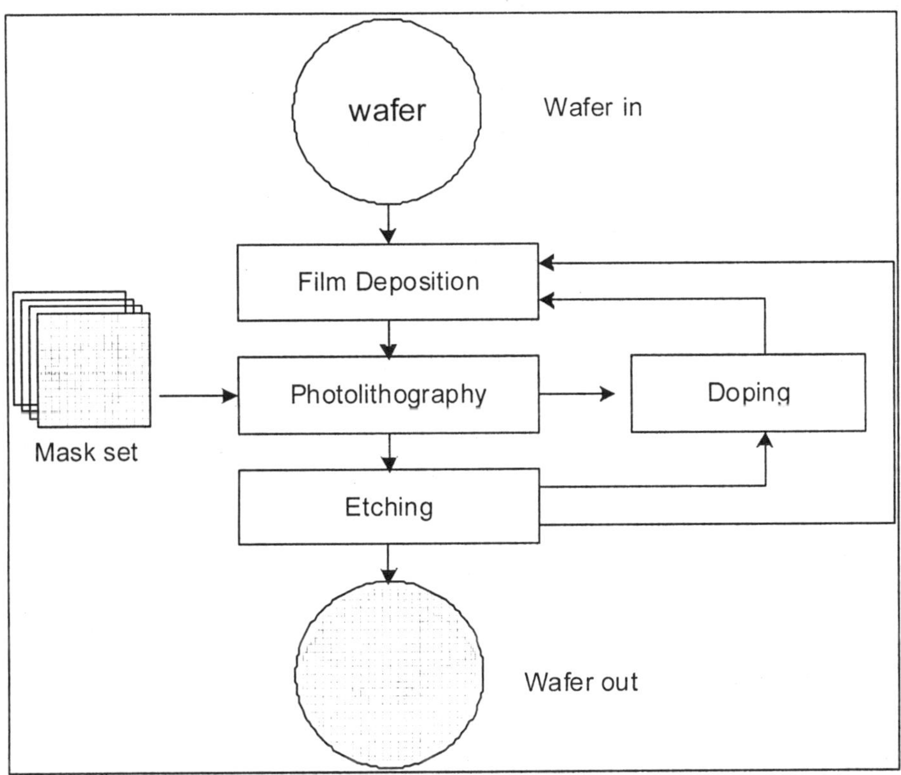

Figure 11.3: Flow Diagram of IC Fabrication Process

Others Materials

This section describes some research results other than those of coordinated research activities by KNRT but have budget supports from the government through the local institution budgets or others collaborations. Other information related to micro/meso/nano-porous materials and chemical self-assembly are being explored in the physical-chemistry laboratory, BATAN. One of the difficulties in inorganic solid-state synthesis is the lack of a small set of techniques for the practitioner to master. The attachment of a single molecular organic layer (self-assembled monolayer, or *SAM*) to the colloidal particles (organic or inorganic) and subsequent self-assembly of those components into a complex structure self assembly is required using molecular recognition and binding.

There has been an enormous growth in the chemical diversity of open framework and nano-porous materials during the last 20 years (Cheetham *et al.*, 1999). Here, the term open-framework refers to materials with open architectures, and use that of nano-porous to refer specifically to materials where porosity has been demonstrated through surface area or ion exchange experiments. Prior to the early 1980s, when the nanoporous aluminum phosphates were first reported by Flanigen and co-workers (Wilson *et al.*, 1982), the aluminosilicate zeolites and closely related systems represented the predominant class of nanoporous materials with three dimensional crystalline structures.

The extent of the change during recent years can be judged from the fact that the transition series are now known to participate as major components of such frameworks and exhibit useful properties that are found with condensed transition metal compounds, *e.g.* ferromagnetism, metallic conductivity, and perhaps even superconductivity that can be further functioned for the design of sensors. Most are based upon oxygen-containing materials, especially phosphates, but there is a growing list of examples based upon other chemistries, such as oxy-fluorides (Fe´rey, 1995), nitrides (Horstmann *et al.*, 1997), sulfides (Bedard *et al.*, 1989; Parise, 1991; Li *et al.*, 1999), and halides (Martin *et al.*, 1997). The range of systems has also increased in another sense: whereas the zeolites and early AlPO4s were based entirely upon corner-sharing tetrahedral.

Other promising synthetic organic/inorganic materials such as pillared clay, glass etc. would exhibit conduction properties that can be functioned as membrane film, conducting based materials etc. Mixed complex compound of silver iodide with silver phosphate glass also has interesting structural, thermal and electrical properties, which exhibits a high ionic conductivity ($>10^{-3}$ S/cm) and can be applied for the in solid electrolytes, batteries, fuel cells and sensors.

Results and Discussion

Quantum Dot

The research groups in the Physics of Electronic Material Laboratory-Physics Department at Bandung Institute of Technology (ITB-Indonesia) performed some research on nano-material electronics. Recently, they performed GaN Quantum dot, which has broad application potency to develop new devices such as single electron transistor and quantum dot laser which can emit short wave length radiation. A laser with quantum dot structure has low current density so that it only requires small power. The growth of GaN quantum dot was done using Plasma-Assisted Metalorganic Chemical Vapor Deposition (PA-MOCVD) method as shown in Figures 11.4a–b. This method is more advanced than that of MOCVD thermal system.

PA-MOCVD System used plasma N_2 which is generated from a microwave source with a frequency of 2, 45 GHz. Metalorganic materials such as trimethylgallium TMGa ($(CH_3)_3Ga$) and trimethylaluminum TMAl [$(CH_3)_3Al$] were used as Ga and Al resources, while tetraethylsilane (TESi) was used as a precursor of wetting layer, where as hydrogen gas was used as a carrier to carry gas vapor from a bubbler. The GaN growth process was done between nitrogen radical generated from nitrogen gas

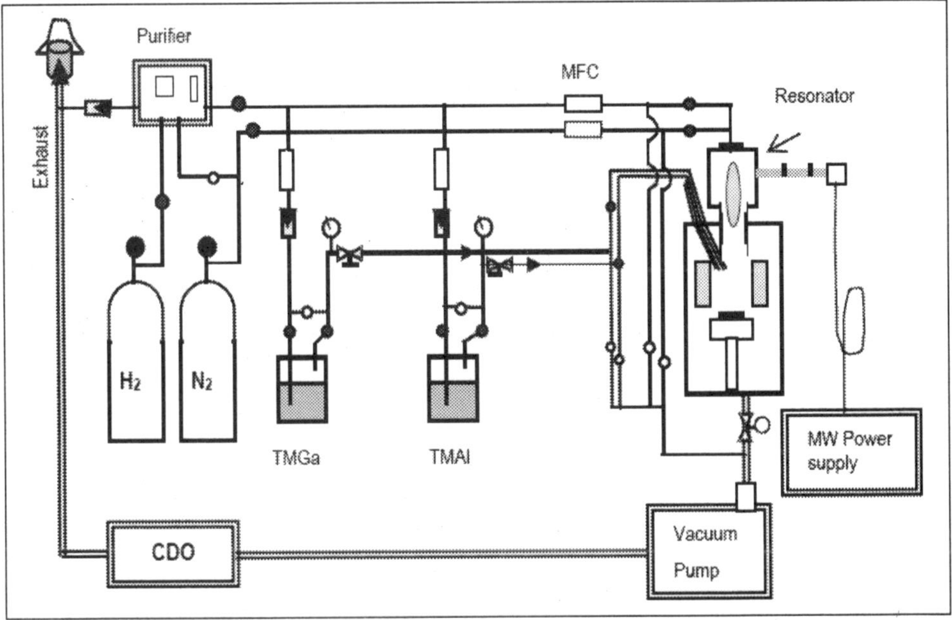

Figure 11.4a: System PA-MOCVD

Figure 11.4b: PA-MOCVD Reactor

plasma with metalorganic vapor TMGa, where as the formation of AlGaN was obtained from the process with TMAl vapor.

$Al_xGa_{1-x}N$ have been made with a certain stoichiometry by TMGa and TMAl flow adjustment using plasma N_2 on the temperature c.a. 600-700ºC. This condition gives advantages on multi layer growth whereas at low temperature the diffusion or evaporation on the film surfaces can be avoided. Whereas without plasma assisted preparation, the process must reach the growth temperature of 1050ºC. The QD growth process was done on 2 types of substrates *i.e.* Si(111) and Al_2O_3(0001). The layer grew firstly at low temperature of about 500ºC with a thickness of 25 nm that is useful to be a layered buffer to avoid possible lattice constants discrepancies between substrate and AlGaN layer. Further process was carried out at temperature of 680-700ºC to grow a layer of AlGaN and the GaN quantum dot occurred with the assistance of wetting layer of tetraethylsilane (TESi) film onto AlGaN layer as shown in Figure 11.5.

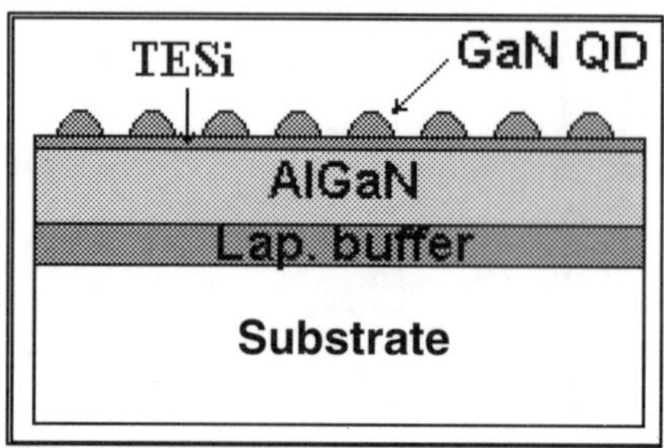

Figure 11.5: Quantum Dot GaN Growth Structure

Figure 11.6 shows topography of GaN QD onto Si (111) obtained by Scanning Electron Microscope (SEM). Dot's size is about 50-100nm with the dot's density of 1 dot per 1.0 μm^{-2}.

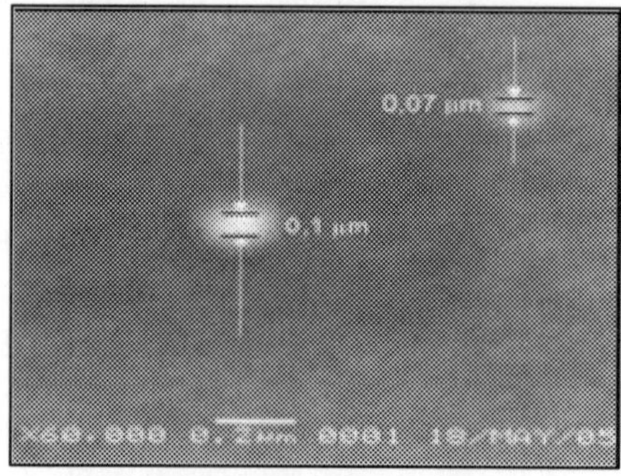

Figure 11.6: Scanning Electron Micrograph of GaN QD onto Si (111)

Figure 11.7 shows topography of GaN QD onto Al_2O_3 (0001) substrate revealed by Scanning Electron Microscope (SEM). Dot's size is increasing by the increasing of time of 20 minutes from about 190–250 nm with the density of 1 dot per 1.0 μm^{-2}, to be 300–400 nm and the density of 1 dot per 3.0 μm^{-2}.

Those were caused by over-growth due to the increasing of time which might have stimulated dots coalition and formed binodal dots with inhomogeneous dots density.

Growth parameter has been found to give a thin layer and produced good result of GaN quantum dots in order of less than 100 nm especially on the Si(111) substrate. Thin layers of AlN and $Al_xGa_{1-x}N$ Al_x have also successfully growth on Si(111) and

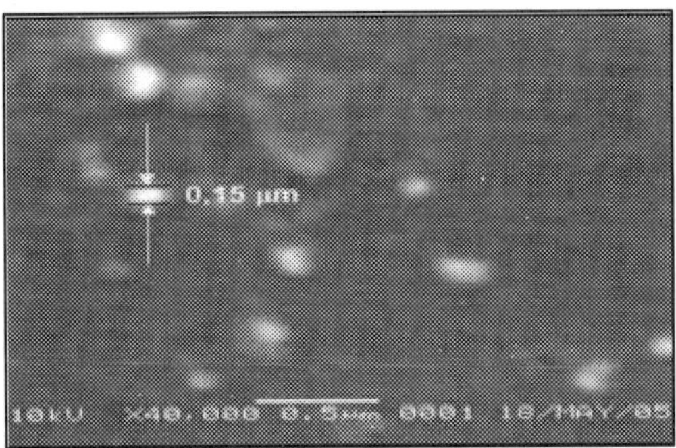

Figure 11.7: Scanning Electron Micrograph of GaN QD onto Al_2O_3 (0001)

Al_2O_3(0001) substrates, which have also fulfilled requirement to form GaN QD device structures.

MEMS: Sensors

Design and fabrication of micro-machined gas sensor has been successfully done by the research group from Research Center for Electronics and Telecommunication, Indonesian Institute for Sciences (PPET-LIPI)-Bandung. Micro machined gas sensor has been fabricated using micro fabrication technology which is initially to be applied to determine carbon monoxide gas (CO). The structure of micro machine is shown in Figure 11.8. It consists of an active area that comprises a Pt-heater, inter-digitated Au electrodes and a gas sensitive layer (tin oxide) situated at the center of a thin membrane, which itself is supported by an outer frame of silicon (micro-heater). The prototype characterization of CO gas sensor was done using the flow cell/chamber to protect the sensor from environmental damage. It consists of vessel tube, gas regulator, flow meter, current/voltage supply, digital multi-meter and thermometer as shown in Figure 11.9.

The synthesizing gas sensor was done in 3 steps *i.e.* using thick film technology of SnO_2 sensing element deposition by spray coating, using thin film technology

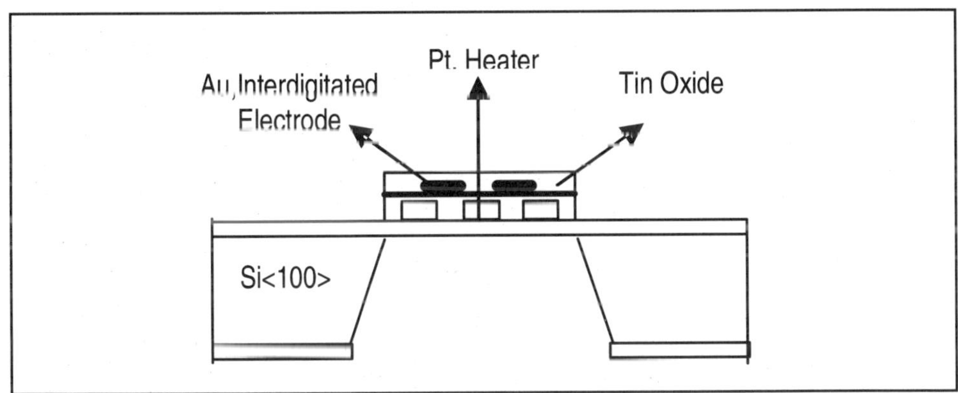

Figure 11.8: Structure of Micro-machined CO Gas Sensor

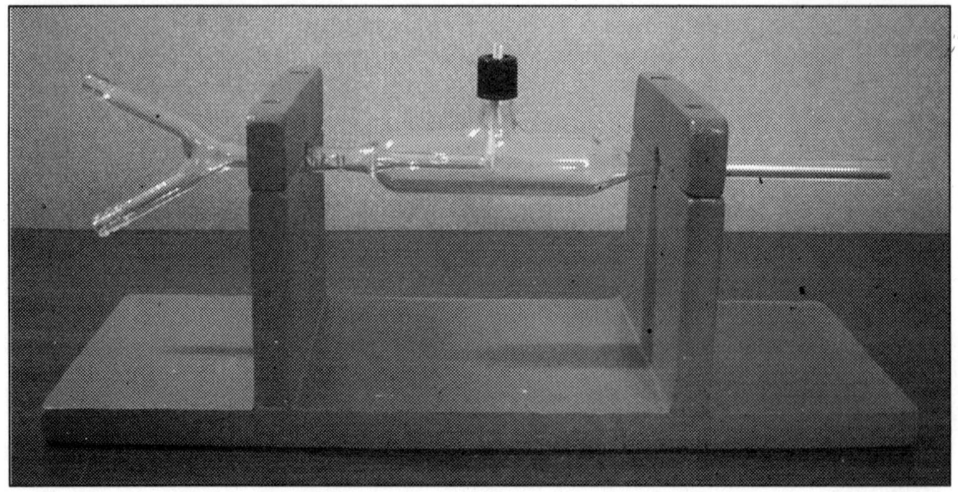

Figure 11.9: Test Chamber to Characterize Gas Sensor

(based MEMS) by anisotropic etching silicon (SiO_2+Si_3N_4 membrane) and using nano technology of SnO_2 particles as sensing element by ballmill agitation. The former has produced a characterized prototype (Figure 11.10) that shown exponential changes of resistance sensor due to increasing of CO gas concentration. This means that at a high concentration the less resistance sensitivity occurs as shown in Figure 11.11. The next and later have not been succeeding to produce good quality sensors and still further studies are necessary to be carried out.

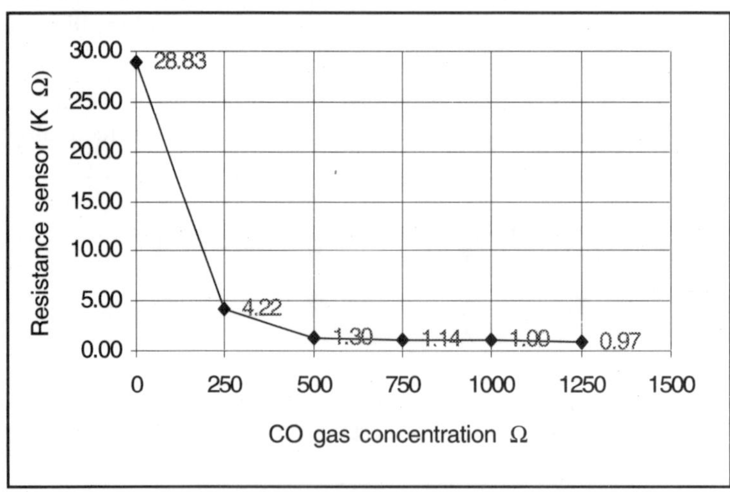

Figure 11.11: Thick Film of SnO_2 Layered CO Gas Sensor Characterization in the Flow Cell. The changes of sensor resistivities (kΩ) vs CO gas concentration.

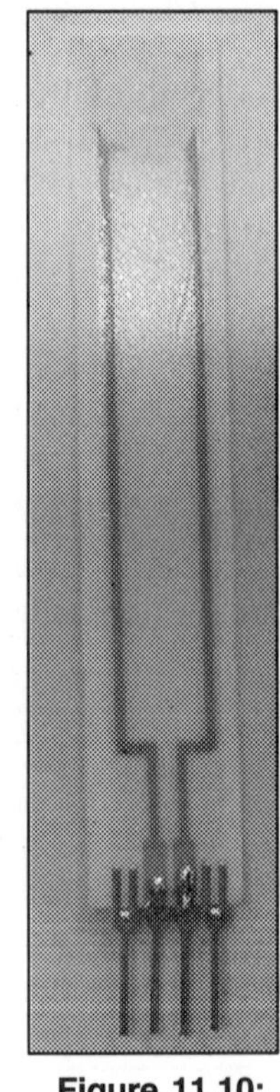

Figure 11.10: Prototype of SnO_2 Layered CO Gas Sensor

Biosensor is also studied to be applied for the determination of glucose concentration in blood or with the other way it can also be applied in others specific biomaterials. The present results just reached the GOD enzyme immobilization onto membrane (cellulose acetate) in the hydrogel system. Three other steps *i.e.* membrane fabrication, reference electrode fabrication and signal conditioning are being in progress completing fabrication and their characterization.

Other Materials and Activities

Silver-based Superionic Conducting Phosphate Glasses, which have interesting structural, thermal and electrical properties, exhibiting a high ionic conductivity ($>10^{-3}$ S/cm) can be applied in the solid electrolytes, batteries, fuel cells and sensors. The research is performed by the material research group from R&D Centre for Industrial Nuclear Material-National Nuclear Energy Agency (BATAN).

On the other hand, many others researchers also performed some researches as per the following information (Indonesian Society for Nanotechnology, 2006) *viz.*:

Dept. of Physics, Institute Technology Bandung exposed the results *e.g.*:

- ☆ I-V meter semiconductor devices use to measure very low current that can also be employed in educational laboratories to find the characteristic curve of resistors, diodes as well as LEDs.
- ☆ Nanostructured catalyst Cu/Zn/Al_2O_3 which is potentially useful for converting methanol into hydrogen in a Direct Methanol Fuel Cell (DMFC).
- ☆ Gadolinium-doped Yttrium Oxide (Y_2O_3: Gd) by employing a polymer heating method. This material is as an ultraviolet source. The particles found to be sub-micron down to several tens of nanometers. The research was supported by the grant of Osaka Gas Foundation of International Culture Exchange.
- ☆ Cerium-doped Yttrium Gadolinium Aluminum Garnet (Y,Gd)AG:Ce by a simple heating method. The yellow green luminescence of nanoparticles (Y,Gd) AG:Ce under the radiation of the blue light is very potential to be used as an alternative solution for the non-vision offline securing system of the bank account book since it is efficient and harmless to our health.

The research group from the Center of Materials Technology, BPPT performed *e.g.*:

- ☆ Synthesis of hydrocarbon membrane for as solid electrolyte in fuel cell.
- ☆ Development of Pt and Pt (Co, Ni, Ru)-alloy catalyst 2-4 nm sizes as electrode material for fuel cell.

The research group from Dept. of Physics, Bogor Institute of Agriculture, and from Faculty of Engineering University of Indonesia has performed some research on optical sensor *e.g.* using of polyaniline nanofiber as modified cladding on fiber optical sensor for ammonia vapor detection, etc., and others topics that can not be describe individually.

Some others related topics with nanoelectronic devices are performed abroad (*e.g.* Japan, Korea, Malaysia and Singapore) as the main tasks of post-graduate students. Other wise, institutions, which have availability of human resources capability, are given in the following table with the following abbreviations:

NM: Nanomaterials,

ND/MEMS: Nanodevice/Micro Electronic-Mechanical System, MEMS,

NT: Nano Tools.

Table 11.1: Summary of Current Status Research and Development of Nanotechnology in Indonesia

Name of the Institute	*Groups*	*Fields of Studies*		
		NM	*ND/MEMS*	*NT*
Indonesian Institute of Sciences (LIPI)	Research Center for Physics	X		
	Research Center for Chemistry	X		
	Research Center for Calibration, Instrumentation and Metrology	X	X	
	Research Center for Electronics and Telecommunication	X	X	
	Research Center for Metallurgy	X		
	Research Center for Electrical Power and Mechatronics	X	X	
National Nuclear Energy Agency for Indonesia (BATAN)	Center for Nuclear Fuel Technology	X		
	Center for Isotopes Application and Radiation Technology	X		
	Center for Nuclear Industrial Materials Technology	X	X	
Agency for Assessment and Application of Technology (BPPT)	Center of Polymer Technology	X		
	Center of Materials Technology	X	X	X
National Aeronautics and Space (LAPAN)	Center for Aeronautics and Space Technology		X	
University of Indonesia (UI)	Department of Metallurgy and Material Engineering	X		
	Department of Electronics Engineering		X	
	Department of Physics	X		X
Bandung Institute of Technology (ITB)	Department of Physical Engineering	X		X
	Department of Physics	X	X	
	Department of Chemistry	X		
Gajah Mada University (UGM)	Department of Chemical Engineering	X		X
	Department of Industrial and Mechanical Engineering	X	X	
	Department of Physics	X	X	
	Department of Chemistry	X		X
Sepuluh Institute of Technology (ITS), Surabaya	Department of Material Engineering	X		
	Department of Chemistry	X		

Contd...

Table 11.1–Contd...

Name of the Institute	*Groups*	*Fields of Studies*		
		NM	*ND/MEMS*	*NT*
Andalas University (UNAND)	Department of Chemical Engineering	X		
Soedirman University (UNSUD)	Department of Physics	X		
Padjadjaran University (UNPAD)	Chemical Engineering Department		X	

Conclusions

Our capability in micro-nano sciences and technology R&D ~~is~~ no doubt has to be strengthened and focused on the available resources subject to the fields of electronics and its related materials.

It is needed to make focused research groups which consist of researchers and nanotechnology players from several institutions to overcome the shortage of human resources, facilities, tools and equipments and also research funds. Also collaboration with other countries is strongly needed.

Information center for nanotechnology should be made through newsletter and internet (website, mailing list etc) for spreading and communicating intensively all information related to nanotechnology activities and research results among researchers and industries.

The field of nanoporous materials has emerged as one of the most exciting and diverse areas of nanomaterials. The discoveries of the last decade have demonstrated that, far from being limited to aluminosilicates and closely related materials, open-framework solids can be made with a rich diversity of chemistry.

The developments in the transition metal area, for example, mean that it is now possible to contemplate the creation of nanoporous solids that exhibit some of the interesting and useful properties that are found with condensed transition metal compounds, *e.g.* ferromagnetism, metallic conductivity, and perhaps even superconductivity.

It is also recognized that pore formation is important in the functionality of ion channel proteins and that proteins can be further functionalized for the design of sensors. This is undoubtedly a very rich field in which many other exciting discoveries will be made in the future.

Acknowledgements

The author gratefully acknowledges many friends and colleagues whose input helped shape this article. Particularly, we acknowledge Hiskia Sirait, Pepen Arifin also other individuals Indonesian researchers in fields of micro/nano materials and electronics for many, many helpful and unflagging discussions. We also acknowledge The State Ministry of Research and Technology and the National Nuclear Energy

Agency for many of the same services, performed in the office of friends. We acknowledge the support services that provided valuable assistance and insights into the literature on capillary formation.

References

Andreas, H., Oliver, B., Christoph, H., Henry, B., 2003. Microfabrication Techniques for Chemical/Biosensors, Proceeding of the IEEE, Vol. 91, No. 6, June 2003. pp. 839-863.

Bedard, R.L., Wilson, S. T., Vail, L. D., 1989. Stud. Surf. Sci. Catal., 49A, pp.375.

Cheetham, A. K., Ferey, G., Loiseau, T., 1999. Angew. Chem. Intl. Ed. Engl., 38, 4000; Harrison W. T. A, 2002. Curr. Opin. Solid State Mater. Sci., 6, pp.407.

Depsey, E., Diamond, D., Smyth, M. R., Urban, G., Jobst, G., Isabella, Moser, Verpoorte, E. M. J., Manz, A., Widmer, H. M., Rabenstein, K., Freaney, R., 1997. Design and development of a miniaturised total chemical analysis system for on-line lactate and glucose monitoring in biological samples, Analytica Chimica Acta, Vol. 346, pp. 341-349.

Ferey, G., 1995. J. Fluorine Chem., 72, pp. 187; C. R. Acad. Sci. Se´ries C 1998, 1, pp. 1.

Horstmann, S., Irran, E., Schnick, W., 1997. Angew. Chem. Int. Ed. Engl., 1992, pp.36.

Indonesian Society for Nanotechnology, September 2006. Current Status of Nanotechnology in Indonesia, Indonesian Nano Letter Vo.1 No.1, Tangerang, Indonesia.

Kovacs, G. T. A., Maluf, N. I. K., Petersen, A., 1998. Bulk micromachining of silicon, Proc. IEEE 86 (8) pp.1536–1551.

Li, H., Laine, A., O'Keefe, M., 1999., Science, 283, pp.1145.

Martin, J. D., Greenwood, K. B., 1997. Angew. Chem. Int. Ed. Engl., 36, pp.2072.

Morcoc, H., Reshchikov, MA., Jones, KM., Yun, F., Visconti, P., Nathan, MI., Molnar, RJ., 2001. Growth and Investigation of GaN/AlN Quantum Dots. Mat. Res. Soc. Symp. 639, GII.2.1.

Parise, J. B., 1991. Science, 251, pp.293.

Pierre, M. *et al.*, 2001. Physics Today, May Vol.46.

Ramvall, P., Philippe, Riblel, Shintaro, Nomura, Yoshinobu, Aoyagi, 2000. Optical properties of GaN quantum dots, J. Appl. Phys. 87, pp.3883.

Shen, X., Tanaka, S., Iwai, Sohachi, Aoyagi, Yoshinobu, 1998. The formation of GaN dots on AlxGa1-xN using Si in gas-source molecular beam epitaxy, Appl. Phys. Lett. 72, pp.344.

Tanaka, S., Iwai, S., Aoyagi, Y., 1996. Self assembling GaN quantum dots on AlxGa1-xN surfaces using a surfactant, Apply. Phys. Lett. 69, pp. 4096.

Wilson, S. T., Lok, B. M., Messina, C. A., 1982. J. Am. Chem. Soc., 104, pp.1146.

Chapter 12

Applications of Micro and Nanoelectronics and Photonics in Kenya

Edith Roseline Njeru

Kenya Broadcasting Corporation, P.O. Box 12664 00100 GPO, Nairobi-Kenya

E-mail: njeru@swiftkenya.com, edith_njeru@hotmail.com, edithrn2006@yahoo.com

ABSTRACT

The Kenyan market saw a marked growth in the use of Computers and Information Technology in the 1990s. Computers are used in virtually all aspects of life in Kenya be it Engineering, Banking, Agriculture, Tourism, Education, Transport etc. Computing devices ranging from mainframes to mini-computers to PCs, Laptops, PDAs to game consoles to consumer electronics devices have been used in huge volumes in all the above sectors. The ways in which computers are being used across the various sectors are Wired networks, Wireless networks, Automotive systems, Consumer and Industrial applications, Television and Radio Broadcasting among others. This paper deals with the various applications of Microelectronics, Photonics and Nanoelectronics in the country.

Keywords: Microelectronics, Photonics, Nanoelectronics, Computer, Information technology.

Introduction

The Kenyan market is not one that is insensitive to emerging technologies especially as far as use of Computers and Information Technology is concerned and more so in embracing the Digital Economy. As early as 1970s mainframe computers were in use in a number of academic institutions and government offices. The 1990s saw a marked growth in the use of computer technology especially with the introduction of Internet in the country. With the possibility to send and receive emails many organizations embraced the use of computers in their applications and a tremendous demand for more and more processing of applications has seen the increase in number of computer devices in the country. Even before computers became common the use of microprocessor based equipment was evident in control systems of various industries, power generating plants, hospitals and academic institutions.

Applications of Micro–Electronics

Computers are used in virtually all aspects of life in Kenya. These include Engineering, Banking, Agriculture, Tourism, Education, Transport etc. Computing devices have evolved from mainframes to mini-computers to PCs, Laptops, PDAs, game consoles and consumer electronics devices. There is increasing demand for more of these devices as the adoption of multimedia content consumption continues to gain ground and be appreciated in many day to day applications. From importing computers wholesome to importing parts for generic clones, Kenyans have recently managed to assemble their own PCs at an affordable cost to the local market.

Pentium IV processor based computers are the ones currently in use for Corporate/office and home PCs. Branded computers (IBM, Compaq, Dell, Toshiba) as well as clones have been in use for a long time. The processors are mainly Intel, Duron, Zylog to name but a few.

Some of the ways in which computers are being used across the various sectors are in:

Networking

Wired Networks

VLSI products and solutions are applied to LANs, MANs and WANs in form of chips encoded in Network devices like hubs, switches and routers. High speed, high performance networks employing intelligent switching devices for Ethernet are employed thus lowering cost of connectivity for applications in data centers. Email, VOIP, data casting, video conferencing and a myriad other applications are all possible through the Local area networks, Virtual private networks (interconnecting a company's offices), Metropolitan Area networks; and Wide Area networks are spanning the country and beyond.

Robust gateway platforms are applied to meet the needs of corporate as well as home networking applications. Academic institutions, banks and other organizations have been able to set up branch to remote office networks for their applications.

Currently digital centres are being set up in rural villages for dissemination of government processes, enabling of e-commerce applications and setting up communication facilities to all parts of the country.

Wireless Networks

Semiconductor products and solutions are being applied for wireless applications to provide a wide variety of current and emerging wireless applications, including those in broadband wireless access, home networking, mobile and storage-network markets.

Wireless networking provides a faster solution for networking to hitherto unreached areas where no communication infrastructure has existed in the past.

WiFi and WiMax technologies as well as many flavours of broadband wireless are entering the Kenyan market at a fast pace due to their ease of deployment and support. Various chipsets are used *e.g.* Atheros, prism2, Orinoco, prism54, broadcom and Ralink.

Automotive

Microcontrollers, graphics display controllers etc are applied for many automotive applications like in fault diagnosis, wheel balancing and alignment, car alarms, car tracking and theft detection. Others are dashboard instrumentation, airbags and other body-control applications. Also in GPS applications.

Consumer Market

There are many technological solutions in appliances for the consumer market. These are in PCs, audio-visual equipment, telephony and general household appliances. A/V chipsets like MP3 decoders, and MPEG encoders/decoders, are used for high-performance audio in VCDs and DVDs.

Digital cameras, mobile phones, TVs and other devices are all in use. Majority of the devices can be used as stand alone or connected to the Internet *e.g.* use of mobile handsets for internet connectivity. Portable computers in form of PDAs and mobile headsets incorporating PC applications are in use. People are able to publish, share, view and listen to content through the Internet from anywhere, anytime. Appreciation of portable, nomadic and mobile devices has increased in the last few years. Almost everyone now owns a mobile set, young people especially students use iPods, MP3/4s and memory cards at a great scale.

Industrial Sector

Advanced technological solutions in ID authentication, security and system appliances for the industrial market are employed. More and more industrial products have incorporated microcontrollers. Large production plants in metal industries, cable manufacturers, cement, plastics and consumer products employ microcontrollers in both their control and production processes.

This enables bulk processing of products in a semi automated environment. The Power generating company also use microcontrollers for Hydroelectric and geothermal power generation in the control and remote monitoring systems.

Magnetic sensors for ID authentication in banks and other sensitive organizations meet the needs of major applications including physical-access systems for buildings, point-of-sale terminals, transaction security systems for online banking, and other e-commerce operations.

Television and Radio Broadcasting Companies

Solid state transmitters employ micro-controllers in their Main Control unit (MCU) for control and monitoring of the signal generation, processing and transmission operations. It is possible carry out fault diagnostics, display and print parameters of the various transmitter modules using the LCD diplay on MCU. Switching on and continuous running/functioning are also controlled from the microprocessor circuit. Other functions include remote monitoring via a web based server interface where parameters can be monitored and transferred via an IP or GSM network.

Studio operations like non linear editing of material, recording and playout are also facilitated through micro-processor based devices. Electronic news gathering, content delivery and storage all use micro-electronic devices in form of Digital cameras, video and audio storage devices, networked applications and field to studio links. Satellite equipment for uplink and down link of signals to remote sites, studio-to-transmitters links are microprocessor based applications.

These are just a few of the applications of Micro-electronics.

Applications of Photonics

Consumer Equipment

Applications in this area include Barcode scanners, printers, CD/DVD devices and remote control devices.

Business

The major use is in Optical detectors in supermarket scanners and other retail outlets.

Medical

Medical optics, correction of poor eyesight and laser surgery all use photonics.

Telecommunications

Here applications are in Optical fiber, lasers and infrared links

Imaging

Weather satellites, night vision, flat screen display, and CCD video cameras

Optical Data Storage and Optical Computing

CD's and DVD's.

Industry

Welding, drilling, cutting, and various kinds of surface modification

Construction

Laser leveling and laser range-finding

Aviation and Military

Command and control, navigation, search and rescue

Applications of Nanoelectronics

The only similarity to Nano technology devices in Kenya are RFID devices commonly used for animal tracking by the Kenya Wildlife Services. Nano-electronics in the real sense is still a technology for the future. It is however expected that this will find a wide range of applications like in clearer displays for Digital TV receivers, health applications, banking, agriculture and rural ICTs.

Conclusion

Microelectronics and photonics have been used in Kenya to a great extent. Nanoelectronics is expected to be adopted soon in various operations.

Chapter 13

Current Status and Future Plan of Microelectronics Technology in Myanmar

Win Win Maw
(Formerly)Department of Electronics, Mandalay Technological University, Myanmar
E-mail:melodywin@gmail.com

ABSTRACT

A brief description of Myanmar, Myanmar Culture and Tradition, Myanmar Education System, Ministry of Science and Technology (MOST), Yangon Technological University and Microelectronics in Myanmar is presented. Myanmar is concerned with the new technologies for ASEAN countries. Microelectronics, photonics and nanoelectronics are closely linked technologies and together they play the most leading part in the utilization of science and technology for development. The high level of technological sophistication is based on the progress made in the fields of microelectronics and information technology. Successive generations of miniaturization of electronic systems have resulted not only in the improvement of their performance but also in the economics of scale. Microelectronics encompasses area as diverse as communication, information technology, data processing and process control to automation of a range of industries as well as biological and medical applications.

Keywords: *Microelectronics, Technology, Myanmar, Photonics, Nanoelectronics, Information technology.*

Introduction

Myanmar is courting domestic and foreign investors in a bid to develop a planned Yadanabon Cyber City in the north. Under the guidance of the Initiative for ASEAN Integration, which is aimed at leveling key social and economic indicators within the regional body, Myanmar will broaden its information and communications technology network. The government also plans to foster domestic hardware production capacity. Authorities claim that the total number of Internet users has reached 300,000. Myanmar is a founder member of the NAM, and Myanmar has been taking active participation in the movement.

Myanmar perceives that the globalization of the international economic system and liberalization of national economics brings both challenges and opportunities to the developing countries. There is general improvement in the economic situation but economic changes have not benefited most people in the developing countries and poverty and economic deprivation have remained rampant in many countries.

Status and Development of World Microelectronic and Nanoelectronics Infrastructure

Microelectronics, which employs miniaturization in electronic devices, is the core technical activity area with immense applications in almost everything. Prosperity and economic future of countries are determined by innovative design, manufacture and ability to use the complex integrated electronic systems. Photonics, which is the science and technology of generating, controlling and detecting photons, has emerged as a separate entity in last 2-3 decades with the invention of the laser and development of optical fibers, photonic crystals, planar wave-guides and other optical elements. It has its applications in light detection, telecommunication, and information demands of photonics products are growing very fast worldwide. Microelectronics, photonics and nanoelectronics are closely linked technologies and together they play the most leading part in the utilization of science and technology for development.

Myanmar is concerned with the new technologies for ASEAN countries. Microelectronics, photonics and nanoelectronics are closely linked technologies and together they play the most leading part in the utilization of science and technology for development. The high level of technological sophistication is based on the progress made in the fields of microelectronics and information technology. Successive generations of miniaturization of electronic systems have resulted not only in the improvement of their performance but also in the economics of scale.

Microelectronics encompasses areas as diverse as communication, information technology, data processing and process control to automation of a range of industries as well as biological and medical applications. Recent awareness of micro-systems has led to opening up of newer areas of application, which were earlier considered outside the realm of microelectronics. The advancements in microelectronics fabrication technology have resulted in the development and fabrication of an assortment of high quality and high precision micro-sensors. Further, low production costs due to the application of batch fabrication technology, small size and low power consumption have facilitated the transformation of the manufacturing of micro-electronics and micro-systems into a commercially viable industry.

Current Status of Myanmar Electronics/Microelectronics Industry

Being an agro-based country, the industries in Myanmar were laying emphasis on development of the production of farm equipments, leading to the establishment of mechanized farming for boosting of agricultural productive forces. It is obviously seen that high technology is developing rapidly and taking an important role in economic and social fields. The range of knowledge and technology is broad and changing and every nation is trying to gain knowledge and technology.

Knowledge, technology and information systems are basic requirements in making the nation strong and powerful. Although Myanmar Electronics/ Microelectronics industry is relatively young, there have emerged various kinds of information and communication electronic apparatus, household and office used electronic apparatus, personal computers, and other electronic apparatus used in factories, hospitals, laboratories, armed forces and spacecraft in various fields. Efforts are thus to be made for speeding up production of high-quality electronic apparatus. Myanmar produces electronic apparatus with three processes — through own design, assembling imported parts, and Chemical-mechanical Planarization (CMP) system. The production sector has not improved as much as it should due to various reasons. However, there are around 250 services, shops or training centers in the electronic sector in Yangon. Some of the established companies that contribute to the electronics/ microelectronics industry in Myanmar are Myanma Machine Tool and Electrical Industries, Hong Pang Group, Earth Industrial (Myanmar) Co Ltd, AA Electronic Ltd, Master Power Co Ltd, Nibban Electric and Electronics, Linyama Electronic Gates Electronics Sale and Service, Mectronic General Electrical Products, Trust and Development Co Ltd, Texcel Co Ltd, U Pe Thein Electrical and Electronic, Toe Electric, Myanmar World Distribution Co Ltd, Fisca Co Ltd, PQ Electrical Engineering Co Ltd and Directorate of Signals. Their products include electronic apparatus–safeguard, transformer, T.V. Antenna, auto voltage regulator for T.V. VCR, refrigerator, air conditioner, electric motor, electric generator, PCB assembly for mobile phone, computer and so on.

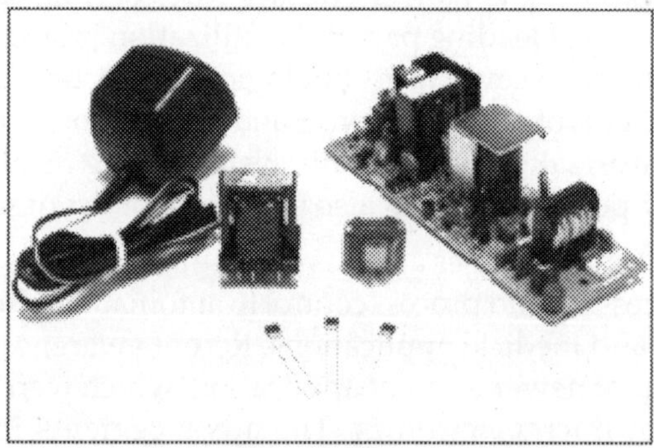

Figure 13.1: Some Electronic Apparatus from Earth Industrial (Myanmar)

The quality system, practiced in the facilities, has been certified to comply with ISO 9001:2000 quality management systems and standards. The products are manufactured to meet strict international safety requirements, for almost all world markets, and test certifications, such as UL, CSA, JQA, DENAN, SGS, BABT, TUV and so on. The organization structure and the range of products of Myanmar

Electronics/Microelectronics industries are more or less similar to that of Myanmar Machine Tools and Electrical Industries.

Organization of Myanmar Machine Tools and Electrical Industries (MTEI)

Managing Director

Planning	*Production & Marketing*	*Finance*	*Administration*
1. Production Planning	1. Production Control	1. Financial Budgeting	1. Administration
2. Production Engineering	and Supervision including Quality Control	2. Financial Management	2. Personnel Affairs
		3. Accounts and Audit	3. Social Welfare
	2. Marketing & Distribution		4. Health and Safety
	3. Materials Control		

Range of Products (From MTEI electronics division)

Electrical and Electronic Factory	*Electrical Equipment Plant*	*Electrical Equipment Plant*
Industrial Zone (1), Dagon Myo Thit (South), Yangon	Inndagaw, Yangon	Sinde, Padaung Township Bago Division

Products	*Products*	*Products*
1. Fluorescent lamps and incandescent bulbs	1. ACSR Cables	1. Electric Generator
2. Electric rice cookers	2. Electric Motors	2. Watt Hour Meters
3. Electric irons	3. Starter Motors	
4. Electric hot plates	4. Dynamo	
5. Dry cell batteries	5. Ignition Coil	
6. Storage batteries	6. Electric Fuel Pump	
7. Household electric accessories		

Future Trend of Myanmar Electronics/Microelectronics Industry

There has been close relationship between the State and private electronic entrepreneurs and experiences, technical know-how and views could be exchanged. Furthermore, the electronic entrepreneurs across the nation have been combined. It is noted that through this combination, an electronic association will be established for elevating the nation's industrial sector, development of electronic industries and manufacturing of international standard electronic products through the strength of experts and technicians from respective ministries and entrepreneurs.

Myanmar has the potential for greater success in producing modern electronic apparatus. Only when Myanmar is able to manufacture modern electronic products, can the nation reduce reliance on foreign countries to a considerable degree in the long run and produce more import-substitution items and penetrate foreign markets. The State owned sector as well as the private sector are to make collaborative efforts for producing quality products with their technical know-how; for emergence of CMP industries in view of producing skilled workers; for setting up joint ventures in view of obtaining technical know-how from abroad; and for undertaking research works to develop their own technical know-how.

Chapter 14

Status of Electronics in Nepal

Nara Raj Giri
Nepal Academy of Science and Technology (NAST),
Khumaltar, Kathmandu, Nepal
E-mail:nararajgiri55@yahoo.com

ABSTRACT

This profile includes the study of the governmental, semi-governmental and some private organizations, which are involved in providing electronics education, electronics product services and consumer electronics goods. This study also shows that there is no major mentionable research and development (R&D) work being done in the field of nanoelectronics and photonics in Nepal. The institutions and organizations which are directly involved in providing electronics education are Tribhuvan University (TU), Purwanchal University and Pokhara University whereas electronics services are being provided by Nepal Telecommunication Company Limited (NTC), United Telecom Limited (UTL), Mero Mobile, Mercantile, Beltronics and LG group. The department of electronics of Pulchowk Engineering College (TU) is providing electronics education up to B.E. and M.Sc. in information and communication engineering degrees. There are some private engineering colleges affiliated to above three universities are also providing electronics education in Nepal. Nepal Telecommunication Company Ltd (NTC) NTC, a semi-governmental organization is engaged in providing telephone services both landline and mobile phones. This organization is using optical fiber in place of usual cables. United Telecom limited, being a private organization is involved in providing wireless local loop (WLL) telephone and mobile phones as well. Mero-mobile being a private company is also involved in providing mobile telephone services. Mercantile, a private company, is providing internet service and supplying other electronics goods for office and household purposes. Beltronics, a private company, is also providing household electronics products and services. The LG, a private Company, is mainly involved in television assembling and other household electronics products. The Ministry of

Environment Science and Technology (MOEST) of the government of Nepal plays a role as a co-coordinator among different national and international organizations and also in the field of science and technology including electronics. Nepal Academy of Science and Technology (NAST) is planning to establish a research laboratory in different branches of modern electronics to provide research facilities to other institutions and concerned individuals. For this NAST is seeking cooperation from SAARC countries and other organizations as well.

Keywords: *Electronics, Organizations, Nanoelectronics, MOEST, NAST.*

Introduction

Nepal is situated in the lap of Himalayas between the latitude 26°22min. N to 30°27min. North and Longitude 80° 12min. East and elevation range from 90min. to 88°48min. East. The average breadth is about 193km. North to South. The country is situated in between the two most populated countries of the world India in the East, South and West and China in the North.

The northern range Himalaya is covered with snow over the year, where the highest peak of the world, the Mount Everest stands. The middle range (hills) is captured by gorgeous mountains, high peaks, hills, valleys and lakes. The southern range (Terai) plain of alluvial soil and consists of dense forest area, national park, wild life reserve and conservation areas. The temperature and rainfall differ from place to place. In the geographic diversity and varying climatic conditions 23.15 million people of more than 60 castes, ethnic groups are accommodated in the country.

Electronics plays a very vital role in the modern civilization. It is the backbone of present technological development. The gradual development in the field of electronics has totally changed the modern life style. The impact of this development is multidimensional in the human life. The development of different branches in modern electronics such as microelectronics, nanoelectronics along with photonics, has greatly helped in the production of electronics goods handy, portable as well as producible in the miniaturized forms which resulted the continue progress in this field has tremendously changed the direction of the world economy. Nepal has also a good market regarding electronics goods, house hold electronics appliances etc. In the context of Nepal even if there is no any major research and development work started in the field of electronics but there are various institutions and organizations which are involved in providing Electronics education, Electronics services, and producing consumer electronics products.

Institutions Involved in Electronics Education

The Pulchowk Engineering College, being the oldest engineering college of Tribhuvan University (T.U) is providing electronics education in Nepal. It has started the bachelors (B.E) degree in Electronics and Telecommunication in 1993. After 2002 this college is offering M. Sc. degree in Information and Communication Engineering also. The Table 14.1 shows the status of electronics education of this college.

Table 14.1: Electronics Education Status of Pulchowk Engineering College

Degrees	*Year of Starting*	*Remarks*
i. M.Sc. Information and Communication Engineering	2002	
ii. B.E Electronics	1995	
iii. B.E Computer Egg	1999	
Branches	i. Electronics	
Electronics Education	ii. Computer Engineering	
Faculty Members	Total No.	
Professors	4	
Readers	5	
Lecturers	27 (Including part time)	
Demonstrators		
(Instructors)	9	
Technicians	3	
Number of students		
Total number in B.E (Including both degrees)	384	
Total number in M.Sc.	32	

Souroo: Office of the Deputy Head of Department of Electronics and computer engineering, IOE, Pulchowk college.

There are some private colleges which are also providing Electronics education. These are Kantipur Engineering College, Kathmandu Engineering College; both are affiliated to Tribhuvan University. Nepal engineering college also involved in providing the electronics education affiliated to Pokhara University.

Electronics Services Providing Organizations

Nepal Telecom NT (Popularly known as Nepal Dursanchar Company Ltd.) is one of the oldest Telecommunication organization of Nepal. It is a semi-governmental organization. It was established by government of Nepal as Telecommunication Corporation. Before 1990AD Nepal Telecommunication Corporation was only one organization involved in telecommunication services. After 1991 government of Nepal changed its policy and allowed the private companies in telecommunication sector. Now the private companies are also working in telecommunication sector and dealing in different types of telecommunication services using different systems.

Table 14.2: Status of Nepal Telecom

Working Manpower	*Total No.*	*Remarks*
Manpower	5701	Technical + Non Technical
Telephone Line Distributed (PSTN)	498784	Landline
CDMA	94753	Wireless
Mobile Phones	704484	Mobile

Contd...

Table 14.2–Contd...

Working Manpower	*Total No.*	*Remarks*
E-mail Service	3686	
Internet Service	5741	
System Applying		
(*a*) Land Line	PSTN	
(*i*) E-10b	SIEMENS (PSTN)	
(*ii*) EWSD	PSTN	
(*iii*) J-RACK	PSTN	
(*iv*) C-DOT	PSTN	
(*b*) Mobile	GSM	
(*c*) WLL	CDMA of ZTE	

Source: Office of Nepal Telecom.

Nepal Telecom also applying optical fiber in place of cable wire along east west highway of Nepal. New Development: Nepal Telecom has developed WC DMA technology in place of 3G system

United Telecom Ltd (UTL)

It is a private telecommunication organization dealing in telecommunication sector in Nepal. It started its working in Nepal 2002. UTL is a joint collaboration of MTNL, TCIC, VSNL and NVPL. This organization has first time introduced the WLL system in Nepal. It is applying fixed WL as well as hand held telephone (HHT). UTL provides telephone services to customer by its 70 staffs.

Table 14.3: Status of UTL

Working Manpower	*Total No.*	*Remarks*
Manpower	70	Tech + Nanotech
Telephone Distributed	50,000	CDMA
Mobile	30,000	CDMA

Source: Office of UTL.

Mercantile

Mercantile also being a private organization offers its services in electronics and related areas from different headings as stated here.

Mercantile Communication Pvt. Ltd.

With this name it provides Internet services to its customers by Internet dial up, broad band, radio link, using optical fiber and VSAT.

Mercantile Solutions Pvt. Ltd

This branch of mercantile deals with different kinds of software. It also develops software as per customer's requirement.

Mercantile Office System

Under this heading, Mercantile deals computer, network equipment, photo copy machines, different types of printer machines and also provides repair and maintenance services to photocopy machines on annual contact basis. It also provides the customer support services outside the Kathmandu valley too.

Beltronics

Beltronics also being a private organization provides electronics services in the areas of office equipment and household consumer electronics goods. The main electronics production of Beltronics is voltage guards, battery chargers, spike suppressors voltage inverters etc. The activities of Beltronics are given below.

Table 14.4: Status of Beltronics Production

Sl.No.	*Type of Product*	*Range*
1.	Voltage Guard	1kV-10kV
2.	Battery Charger	12V, 5A, 48V, 20A and 96V, 20A
3.	Inverter	125VA – 1 kVA
4.	Spike suppressor	1 Line to 4 Line
5.	Voltage Inverter (Step Down)	220V to 110 VA
6.	Isolation Transformer	100W to 500W
7.	Precision power supply (PPS)	5V, 250 mA to 18V, 3A

Source: Beltronics office.

The production of Beltronics is not exported, it has in-house market only. The need of raw materials is fulfilled by import from different countries, mostly from India. Beltronics does not do any mentionable research work in the field of electronics.

Role of Ministry of Environment Science and Technology (MOEST)

The ministry has its role as a coordinator body among different national and international institutions involved in scientific and technological activities. The other important role of the ministry is to formulate the national policy suitable to nation for science and technology.

Role of Nepal Academy of Science and Technology (NAST)

The Nepal Academy of Science and Technology (NAST) is an apex body of science and technology of Nepal. It is actively involved in research and development work in different areas of science and technology. NAST is planning to initiate research work in the area of electronics and modern field of electronics as well. NAST is seeking co-operation from scientific institutions of SAARC and other countries for research work in modern branches of Electronics.

Acknowledgement

I would like to thank NAM S&T Centre, Delhi and COMSATS, Islamabad for providing the opportunity to present this paper in this workshop cum Training program. Similarly, I would like to thank Prof. Dr. H.N. Bhattarai, Vice Chancellor, NAST for his kind cooperation to nominate me to participate in this programme. Finally, I appreciate the support of different organizations and individuals who have provided me the necessary information to prepare this paper.

Chapter 15

SMS (Short Messaging Service) Based Data Transmission System for Automated Rain Gauge Stations in Sri Lanka

Balage Asanga Indunil
Electro Technology Laboratory, Industrial Technical Institute,
No. 363, Bauddhaloka Mawatha, Colombo 00700, Sri Lanka
E-mail: asangais@iti.lk, asangais@yahoo.com

ABSTRACT

An automated, remote rainfall gauging system is developed to monitor the rainfall in five remote areas from one base station. Two replica base stations are going to be installed at the Department of Meteorology, and National Building Research Organization (NBRO). The base stations and the ARGSs are connected with a Global System for Mobile communication (GSM) link. The ARGS sends rainfall data every hour unless an abnormal rainfall (when the intensity of rainfall exceeds a pre-determined threshold level) is detected. The automated rain gauge station consist of a PIC18F452 microcontroller based processing and controlling system, a dual band 900/1800 MHz Fargo Maestro100 GSM module based communication link, DS12887 Real Time Clock (RTC), a lightning protection system, and a surge protection system. Rain gauge is built with a Tipping Bucket, which, is a mechanism used to measure rainfall in an unattended manner. The minimum detection limit of the Tipping Bucket arrangement is 0.25 mm of rainfall with an accuracy of 4 per cent . Rainfall together with date and time obtained from the RTC is sent as an SMS (Short Messaging Service) every hour to the base stations. Current rainfall can be observed by an LCD (Liquid Crystal Display) attached to the ARGS and the backlight of the LCD is automatically illuminated by capturing the presence of human being via IR (Infrared) proximity sensor. The ARGSs are powered by national power grid and equipped with a 12 V rechargeable sealed Lead-Acid battery backup for 5 days in case of a power failure. The base station consists of a PC (Personal Computer) and the Fargo Maestro100 GSM module. A GUI (Graphical User Interface) based software application displays the rainfall individually for all ARGSs in graphical and tabular formats. The captured rainfall data from all ARGSs is saved.

Keywords: *Microcontroller, GSM modem, SMS, Automated rain gauge, Tipping bucket, AT commands.*

Introduction

Collecting rainfall information is a crucial requirement of many areas in science as well as in natural disaster warning systems. The rainfall intensity and soil water level are key factors causing landslides. Rainfall is measured once a day as an average value. Sri Lankan Standard day for rainfall measurement is defined from 8.30 a.m. to the next day 8.30 a.m. Almost all the weather stations located in Sri Lanka are manually operated systems. Manned weather stations have inherent drawbacks. Obviously those require a lot of man power. Rain gauges, like most meteorological instruments, should be placed far enough away from structures, and trees. Due to that reason, it is difficult to obtain readings during rainy or stormy periods. Also the time information is hidden in the conventional method, instead an average rainfall is measured. But for an early disaster warning system, it is essential to have the rainfall with time information. Also it is not realistic to establish a manned weather station, where it is difficult to reach.

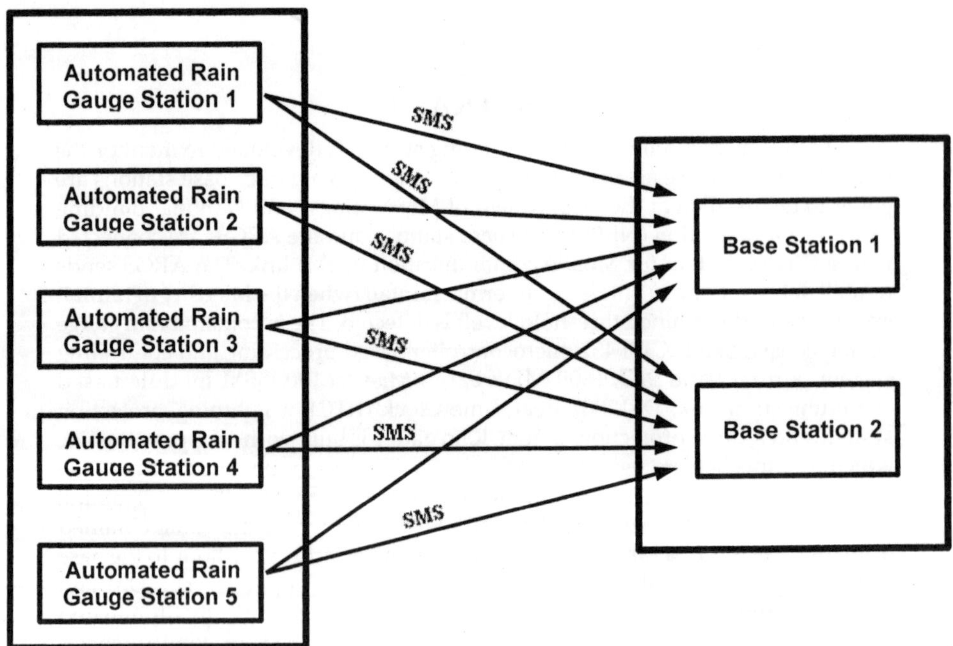

Figure 15.1: ARGSs and Bases Station Communication

With such drawbacks, for an early disaster warning system, manned weather stations are not good solutions. Thus an automated, stand alone, remote operable rainfall gauging system is developed to monitor the rainfall. The automated rain gauge stations (ARGSs) and the base stations are connected with a Global System for Mobile communication (GSM) link. Communication is done through short messaging service (SMS). The ARGS sends time together with rainfall every hour unless an abnormal rainfall (when the intensity of rainfall exceeds a pre-determined threshold level) is detected. In the event that an ARGS detects a high intensity in rainfall, a special warning message with rainfall is sent to base stations as an alert, and rainfall data is sent to the base station every five minutes until the abnormality disappears.

Materials and Methods

The system consists of two components. ARGSs are the on-site operating component and the other component is base station. Data capturing is done by ARGSs, and sent to the base station.

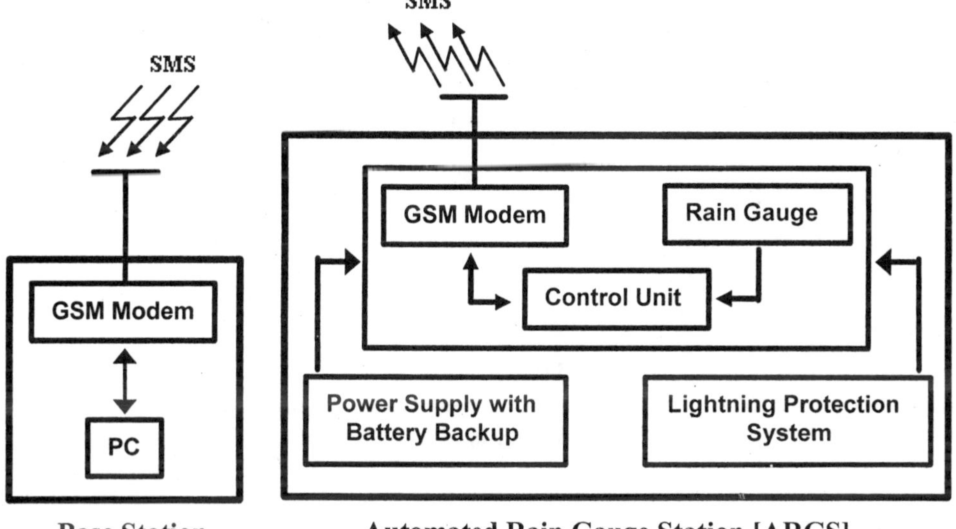

Figure 15.2: Block Diagram of Complete System

Automated Rain Gauge Station Components

In a normal rain gauge, rain is manually collected and most often readings of the gauge are also logged manually. But in automated rain gauge all the data should be acquired and processed automatically. In addition ARGS broadcasts rainfall figures via a GSM link to a base station (Figure 15.3).

The ARGS also determines the intensity of rainfall and change its rate of acquisition accordingly.

In order to achieve these tasks a microcontroller based system is utilized in the ARGS. It's also equipped with a RTC (Real Time Clock), an LCD, an IR transceiver,

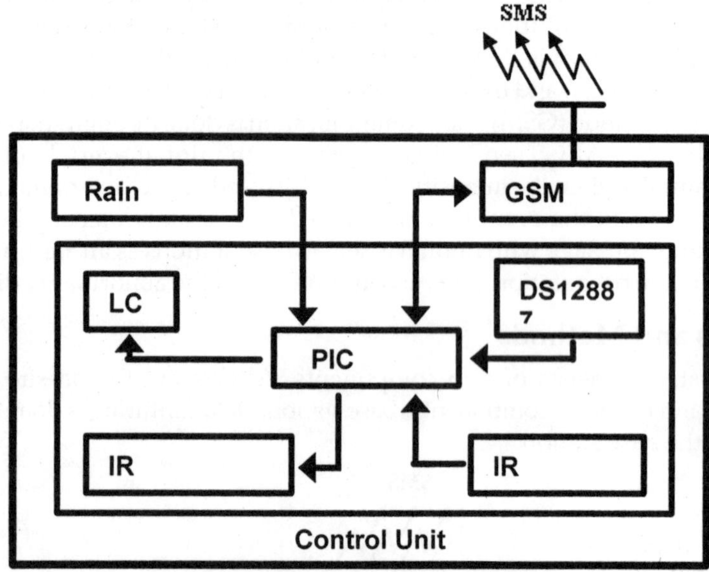

Figure 15.3: Circuitry Block Diagram of ARGS

and emergency battery backup. A lightning and a surge protection system are also utilized. The functionality of these components are discussed below.

Rain Gauge

This is a mechanism developed to measure rainfall in an unattended area. It has a funnel of about 160 mm in diameter contained in an Aluminum jar. Whenever there is rain the water collected in the funnel is allowed to drop to the tipping bucket mechanism. Usually the mechanism tips for every 0.25 mm of rain. The tipping action turns on a magnetic Hall Effect sensor Honeywell SS441A momentarily. By counting the number of times the bucket tips an accurate measurement of rainfall is acquired. The output of the Hall Effect sensor is connected to built-in 16 bit hardware counter of the PIC18F452 microcontroller. This allows a maximum rainfall measurement of 4096 mm per day without overflowing the counter.

GSM modem and SMS generation

SMS alert generation is done with a Fargo Maestro GSM100 modem connected to the microcontroller. The modem supports GSM V 7.05 text messaging specifications. PIC18F452 microcontroller and the modem are connected serially via a RS232 interface. This interface consists of hardware UART (Universal Asynchronous Receiver Transmitter) at the microcontroller and a RS232 transceiver IC (MAX232). Like any other modem a GSM modem is also controlled with a set of text commands called AT commands. At the power on microcontroller sends the necessary commands to detect the modem and to initialize it.

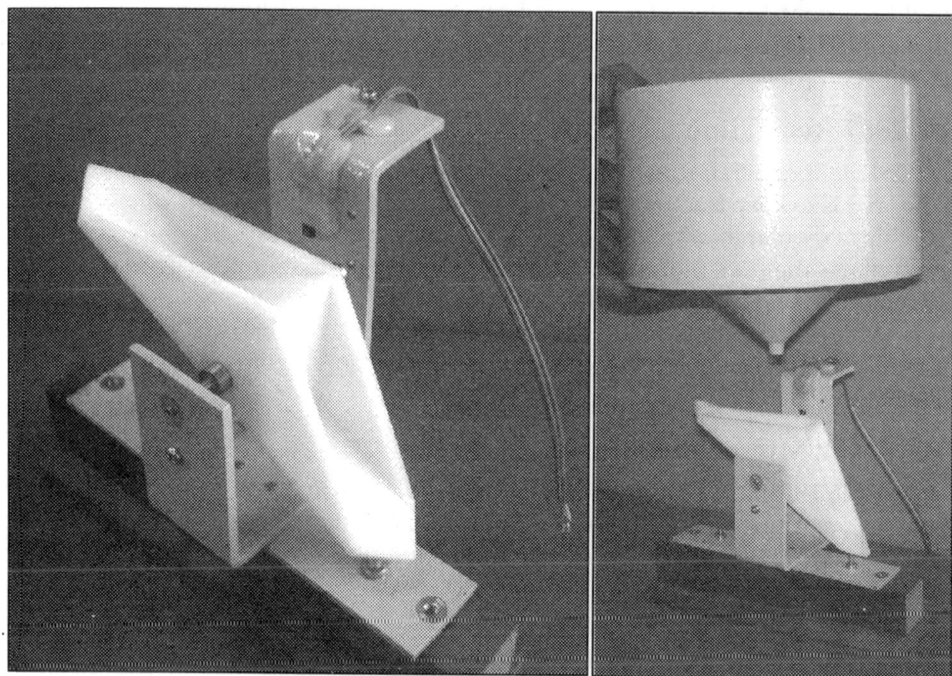

Photograph of the Tipping Bucket of the Rain Gauge **Photograph of Tipping Bucket with the Funnel**

Figure 15.4

Then the microcontroller waits for the GSM modem to get registered with the GSM operator. After a successful network registration microcontroller begins its main function of taking rain gauge measurement and keeping track of time.

Table 15.1: Sample AT Commands and Descriptions

Code	*Description*
START045.25 05-01-2007 09:30STOP	Sample SMS body
AT+CMGS="+94776955737\"\nSTART045.25 05-01-2007 09:30STOP \x1A	AT command
AT+CFUN=1\n\r	AT command to reset
ATE0	AT command for Echo off

The critical factor, when controlling the GSM modem with the PIC18F452 microcontroller, is receiving and processing the responses from the modem. For each AT commands the modem responses with a text string which indicates whether the command was successful or not. These responses should be correctly interpreted and necessary steps should be taken in order to maintain a successful operation. For example sending an SMS might fail due to network conditions and in this case it should be resent. Because of the limited processing power of the microcontroller (10 MIPS at 40 MHz) this is achieved with highly optimized text comparison routines at the microcontroller firmware.

DS12887 Real Time Clock (RTC)

DS12887 is a parallel interfaced RTC chip with built-in lithium battery. Function of the RTC is to keep track of real time. The microcontroller reads the time from the RTC chip at regular intervals of 30 s. This data consist of an array of bytes where each byte contains the year, month, date, day of the week, hour, minute and seconds. The real time obtained with RTC is used to generate SMS alerts of rainfall at every hour. And in the event of intense rainfall SMSs are generated at every 5 minute. Because of the built-in Lithium battery, RTC chip retains correct date and time even at a complete power failure.

PIC18F452 Microcontroller

The PIC18F452 microcontroller is the heart of the ARGS. Acquisition of rainfall figures and real time from RTC, displaying the results in the LCD and most importantly communicating with base stations through the GSM Link is done by the microcontroller.

PIC18F452 is a RISC (Reduced Instruction Set Computer) driven by a 40 MHz clock and this gives an execution speed of 10 MIPS (Million Instructions per Second). The main features of the microcontroller include hardware counters and timers, hardware UARTs, several interrupt sources and five IO (Input Output) ports. Throughout this project these features are utilized in order to achieve the final task.

Microcontroller firmware was developed using a C language compiler called HI-TEC PICC integrated into MPLAB IDE. To program and debug code Microchip ICD2 was used. The complete firmware development life cycle starting from writing code, simulating, programming and In Circuit Debugging was carried out with this development platform.

The final firmware is programmed into the microcontroller using a Boot-loader software making use of PIC18F452's self programming feature. This enables the firmware updates to be done using the serial port of the microcontroller without removing it from the control box. This also enables remote firmware upgrades directly from base stations with few modifications to firmware and Base Station software in the future.

Power Supply

ARGSs are powered by national power grid (230 V AC at 50 Hz) directly. Stepped down voltage is fed to the system through a 12V rechargeable sealed Lead-Acid battery. Therefore small voltage fluctuations are eliminated and the rechargeable battery acts as power backup for 5 days time in case of a power failure and recharging is done automatically. When a power failure is detected, the system moves to low power consuming mode by switching off LCD backlight, and other indicators.

When someone is close to the control box of the ARGS, the backlight of the LCD is automatically lighted and the proximity of personal is detected by the active IR transceiver. This IR proximity sensing system is also utilized to reduce the power consumption.

Power surge, Lightning, and Electro Magnetic Interference Protection

The system is protected from the ground flashes with the help of Franklin type Lightning Rod. A Galvanized Iron rod is dipped up to 240 cm covered with Gemco gel in order to make good contact with the soil to provide earthing to the Franklin type lightning rod of 300 cm high, and is connected to the lightning rod by copper stripes. The metal box, where the system is placed provides Faraday cage protection to the system from external electromagnetic interference and the box is also connected to the ground by copper stripes. All the sensitive electronic circuitry is connected to the box through capacitors to avoid noise. A screened wire was used to communicate with the sensor of the rain gauge and the controlling unit for noise elimination. The wire which supplies electricity to the system also dipped up to 30 cm to avoid generation of power surges by lightning. Further system is protected from the power surges with the help of Micro power surge protector.

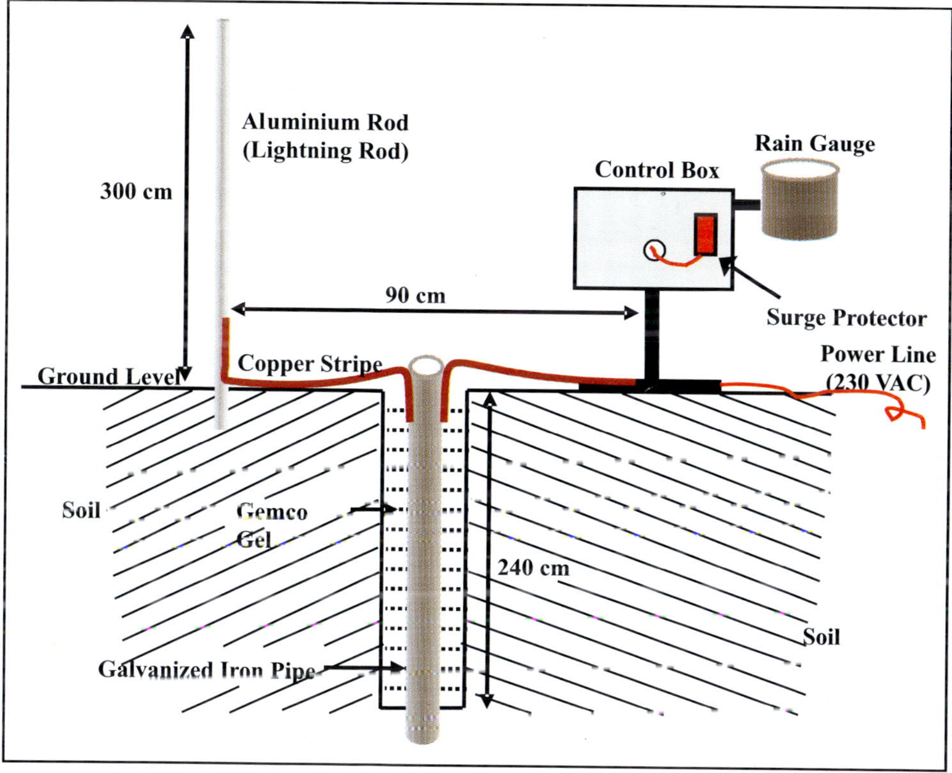

Figure 15.5: Surge Protection System at the ARGS

Base Station and Software

Base station is the data receiver from the automated rain gauge stations (ARGSs). All the ARGSs are directly connected to the base station via GSM link. There are two

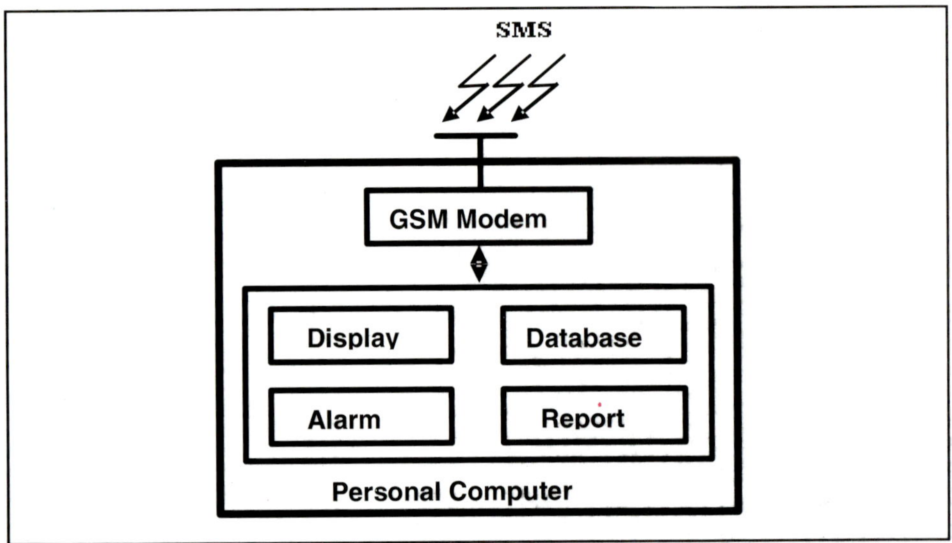

Figure 15.6: Base Station

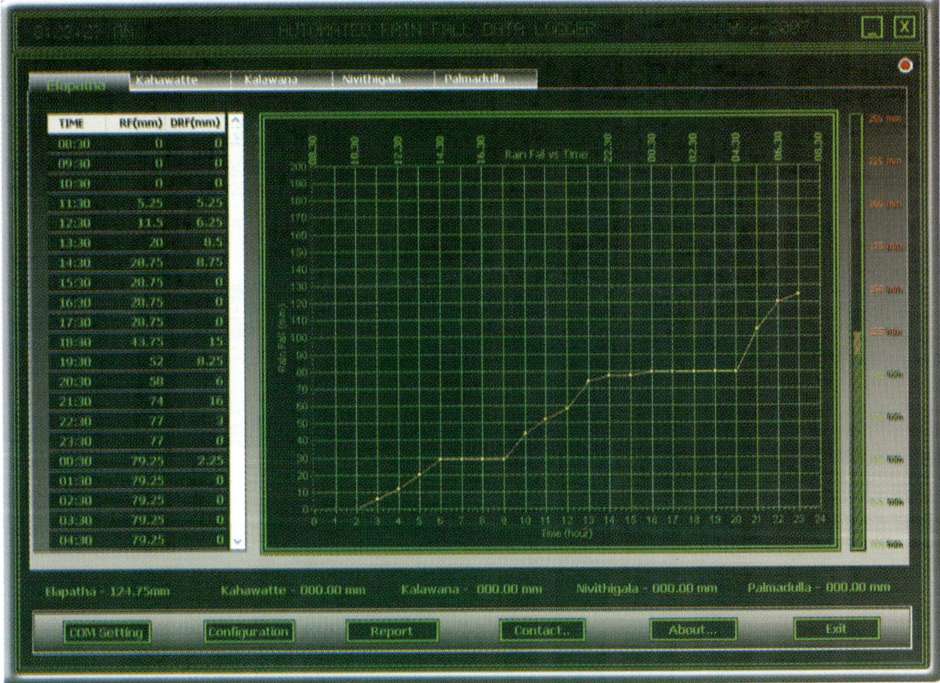

Figure 15.7: GUI of the Software Automated Rainfall Data Logger

replica base stations are utilized in this project. One is installed at the Department of Meteorology, and the other is installed at the National Building Research Organization (NBRO). Both receive same data separately, but simultaneously.

The base station consists of a PC (Personal Computer) running with Microsoft® Windows® XP (SP2) and the Fargo Maestro100 GSM module. A GUI (Graphical User Interface) based application software application developed using Borland® Delphi 6. The application software at the PC end is the only part of the design which interacts with the user. Therefore every effort has been made to make this part efficient, user friendly and reliable. The application software is given the name "Automated Rainfall Data Logger".

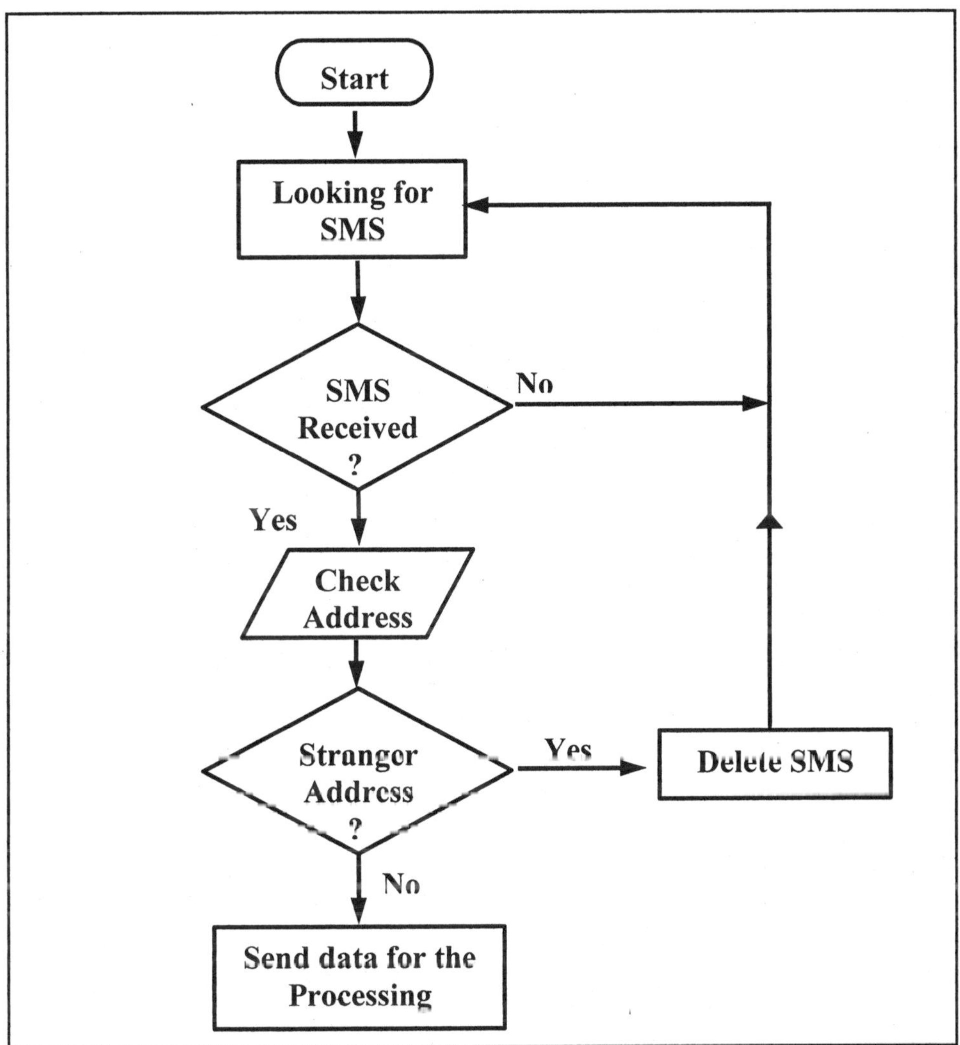

Figure 15.8: Algorithm of Receiving Data to a Base Station

GSM modem is connected to the computer via the serial port (COM1). The data at the com1 port is captured by a third party Delphi component called "CPORTLib" integrated to the software. When software initializing, GSM modem is registered in the GSM network. When registration gets succeeded, flashing indicator will display on the GUI of the software.

Each ARGS has unique GSM address (or number). The software uses this GSM address (or number) to identify the ARGS which sent the data. Since this is a general GSM modem it is possible to receive SMS from some other phones as well. If an SMS from a strange number is detected, the SMS will be automatically refused, and deleted by the software. If data from an ARGS is detected it will be recorded by the software.

Following is a sample data from an ARGS. +CMT is a controlling signal generated by the modem, and the next +94777958112 is the GSM address of the ARGS, followed by the date and the time and then in between START and STOP is the data rainfall, date, and time.

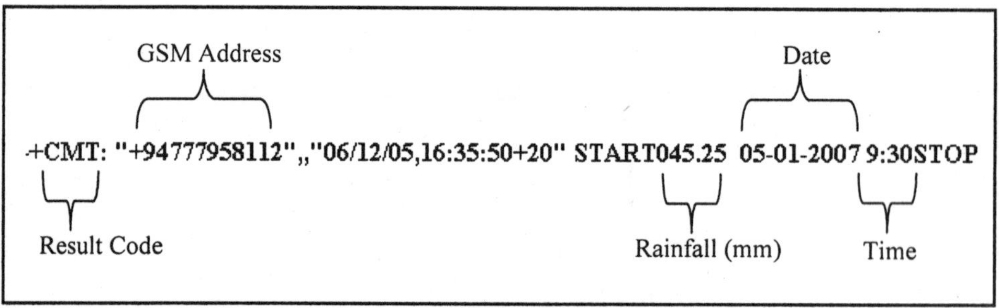

Figure 15.9: Sample Data from an ARGS

The data from individual ARGS is tabulated in separate tables and the graphs are automatically updated. The software allows user to retrieve data in tabular format as well as graphical format for individual ARGS. Cumulative rainfalls for all ARGSs are always displayed in the status bar of the GUI. At the end of each day, the data is saved as CSV (comma separated value) file (which can be opened through almost all the spreadsheet application). A new folder is created for each day and separate CSV files are created within that folder for each ARGS. A Sri Lankan Standard day for rainfall is defined from 8.30 a.m. to the next day 8.30 a.m. Software starts a new day at 8.30 a.m.

Result and Discussion

Five ARGSs were installed at five selected locations in Ratnapura district of Sri Lanka. (Ratnapura district is the most vulnerable for landslides in Sri Lanka). One base station is installed at Industrial Technology Institute (ITI) for testing purpose and it functions well. Two replica base stations are going to be installed at the Department of Meteorology, and National Building Research Organization (NBRO). Final objective of this project is to develop computerized early warring system to predict landslide possibilities using rainfall and ground water level value as input data. While operating under normal conditions, the ARGS has a typical power consumption of 0.5 W.

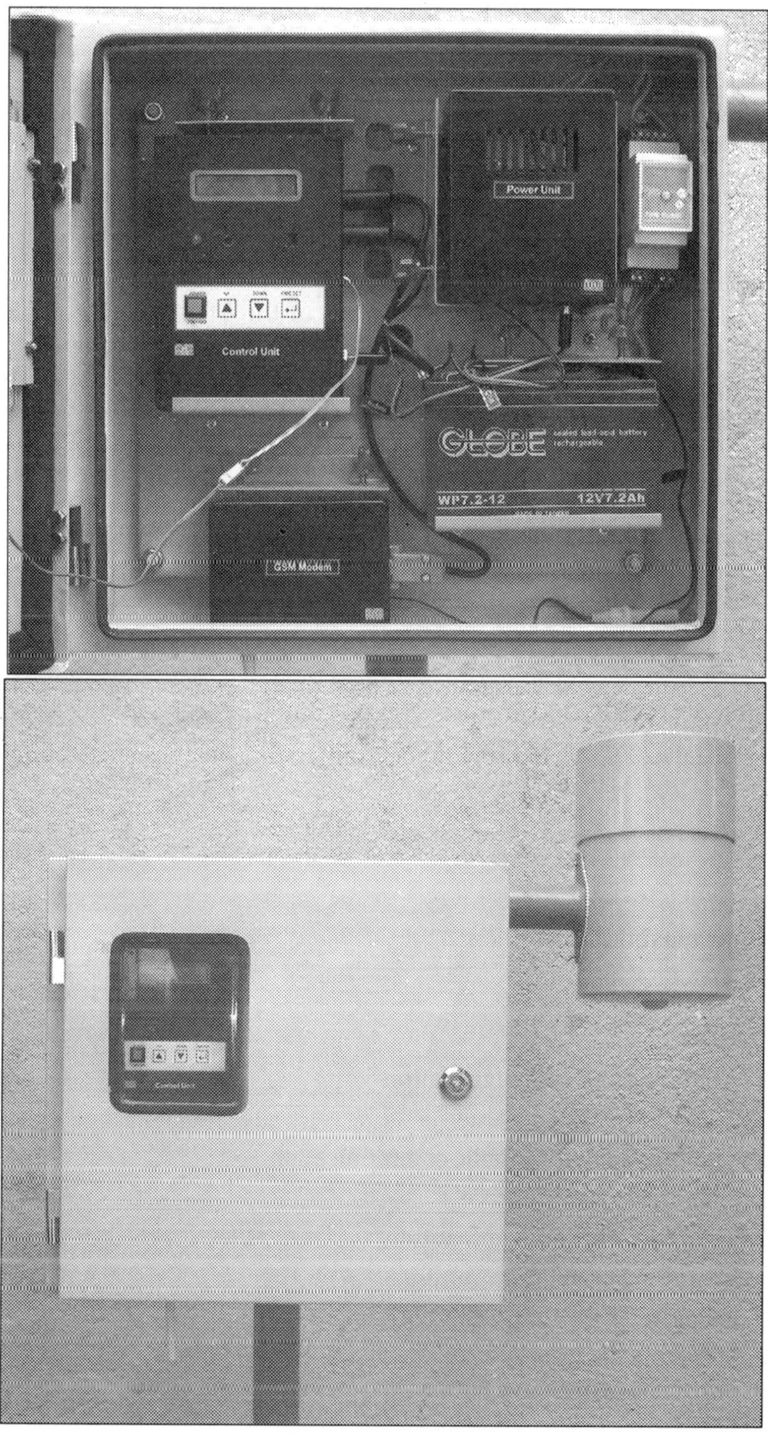

Figure 15.10: Top: Photograph of the Opened Control Box at ARGS; Bottom: Photograph of Full System at the ARGS (Control box is closed)

Energy of the fully charged battery is 12 x 7 AVh. The battery allows running the system without any failure for 5 days time. Also if there is a power failure in the base station. No data will be lost, because SMS are stored in the servers of the GSM service provider. As soon as power is resumed, all the SMS will be downloaded to the PC (Personal Computer) located at the base station.

The hardware requirements of the base station are an IBM PC or compatible computer with a processor equivalent to Pentium III 400 or better, serial (com) port, the recommended RAM for the operating system used. The PC should have Microsoft Windows 98, 98SE, 2000, ME or Windows XP operating system installed. The ARGS makes full use of special peripheral features of the PIC-18F452 microcontroller. Features such as hardware timers, UART, ICSP (In-circuit Serial Programming), Self Programming and interrupts are fully utilized in this project. An ARGS must be located in an area where GSM network coverage is present. Sometimes there are delays in sending/receiving SMS in current GSM networks in Sri Lanka. But it can be easily eliminated by taking priority for the GSM numbers used for ARGSs from the GSM service providers. Also the cost for communication can be reduced by using GPRS as data bearer rather than SMS. Although there is a good lightning protection system is employed, still there is a threat of lightning striking to the system.

Bibliography

Heath, Steve, 2003, "Embedded systems design", Newnes, Oxford.

Iovine, John, 2000, "PIC microcontroller project book", McGraw-Hill, New York.

Sommervile, Ian, 2001, "Software Engineering", Addison-Wesley, Reading.

Websites

http://www.microchipc.com

http://www.microchip.com

http://www.maxim-ic.com

http://www.wavecom.com

Abbreviations

ARGS: Automated Rain Gauge Station
CSV: Comma Separated Value
GSM: Global System for Mobile communication
GUI: Graphical User Interface
ITI: Industrial Technology Institute
ICSP: In Circuit Serial Programming
IDE: Integrated Development Environment
ICD: In Circuit Debugger
IR: Infrared
IO: Input Output

LCD: Liquid Crystal Display
MIPS: Million Instructions per Second
NBRO: National Building Research Organization
PC: Personal Computer
RTC: Real Time Clock
RAM: Random Access Memory
RISC: Reduced Instruction Set Computer
SP2: Service Pack 2
SMS: Short Messaging Service
UART: Universal Asynchronous Receiver Transmitter

Chapter 16

The Teaching of Microelectronics and the Prospects of Research Activities in Tanzania

Edward E. Mhamilawa
Department of Electronics and Communication, University of Dar es Salaam, Dar es Salaam, Tanzania
E-mail: nusesa-tz@udsm.ac.tz

ABSTRACT

The Tanzania Commission for Science and Technology (COSTECH) was established to promote science and technology in the country through research supports of Research and Development (R&D) institutions. Electronics is among many scientific research fields being pursued in several R&D institutions. Most of the reported research activities done in faculties of science, engineering and many other technical institutes are in the application of electronic devices. There are few postgraduates who studied abroad and worked in advanced laboratories to study electronic materials and microchip fabrication. These graduates are the hope in forming the nucleus of research activities in the microelectronics in the country. A survey of existing electronics research laboratories in the country show that they cannot manage to support research in micro-and nanoelectronics field. There are great hopes towards better future in education and research in Tanzania. The results are seen in the astronomical increase of universities in the country. As an example, from the year 2000 to 2007 the number of universities has increased from four to twenty two and more are under construction. This shows the potential of growth of research activities, both in quality and quantity. The University of Dar es Salaam is the leading institution in science and technology research fields. The establishment of a good research laboratory for the microchip fabrication is an expensive affair. However, with the availability of researchers, the positive attitude of the government towards education and research and the current revolution in the industrial sector it is a matter of near future to see microelectronics researches established in Tanzania.

Keywords: *COSTECH, MHEST, Micro-electronics, Nanoelectronics, University of Dar es Salaam.*

Introduction

Tanzania, a non-aligned East African member state, does have a Ministry of Higher Education and Science and Technology (MHEST) tasked to administer advanced educational and research activities in the country. The ministry was introduced for the first time in 1990 after realizing the importance of Science and Technology in the country's development. Prior to this ministry there was a Commission for Science and Technology (COSTECH), established by the Government Act of 1986. It was and has remained to be the principal advisory organ to the government on all matters related to scientific research and technology development in the country. It has the task to coordinate and monitor various research supports in R&D institutions and to promote science and technology in the country. The COSTECH is a brain child of the Tanzania National Scientific Research Council, an organ launched in 1972 for the same fundamental objective.

There are many research fields under the administration of COSTECH. These include researches in food and agriculture, natural resources, environment, medicine, basic sciences, social sciences, industry and energy. Whereas all these fields are interrelated but it is best to note that they all use technologies developed from basic science research. When one thinks of instrumentation and its use in research activities then electronics comes to play. Electronics is one specific scientific research field conducted in few Tanzania R&D institutions, typically at faculties of science; engineering and other technical institutes working with electronic devices. These electronic devices are fabricated with the help of the microelectronics knowledge; a key to the understanding of the world of electronics that controls the evolution of instrumentation that in turn control the development of science and technology. The miniaturization of electronic devices has gradually moved from micro- to nano-dimensions and the application of photonics technology is growing fast. Thus the knowledge and understanding of the design and fabrication of the devices demand more involvement in the microelectronics research field. In a third world country, like Tanzania, the strength of microelectronics base lies on the industrial demand of electronics that use the trained electronic engineers and the scientists working in the education system. Any success story that boasts on the quality and quantity of microelectronics activities in teaching and research in R&Ds depend highly on the existing infrastructure and the financial support apportioned by the government to the education and research sector. It is therefore important to know the national policies that support the growth of science and technology that can give a leeway to the development of microelectronics research fields. It is also important to know the growth and trends that support science education at secondary, university and tertiary levels. Future activities in fields of microelectronics are gauged by the current investment in these fields in terms of financial and human resources and the will of the government to support the activities. The relevance of these research activities to the nation must be sighted in the national goals and must be in line with the international goals ratified by the government, such as the OAU Lagos Plan of Action1980-2000 and the UN Millennium Development goals.

Science and Technology Education

The success of achieving the basic goal of doing research in microelectronics: micro- and nano- electronics and photonics in Tanzania lie on the current situation of infrastructure, human and financial resources. A brief look at the education system that outputs the human resources and its developmental trend can throw some light on the possible future of microelectronics research activities.

For a very long time, Tanzania had a very slow university education expansion. This included the lack of establishing many more national and private universities in the early years of its independence (1961). Similarly, the expansion of student intake in the existing universities remained very low. Chronologically, Tanzania had its first university college in 1961 as part of the university of East Africa and received its autonomy in 1970 and became to be known as the University of Dar es Salaam and it remained to be the only full fledged University in the country for 14 years. In 1984 the Faculty of Agriculture of the University of Dar es Salaam became a full fledged University of Agriculture, to become the second university in the country. In 1992 the Tanzania Open University was born to mark the third public university. During all this time the private sector did not join in the expansion of university education until 2000 when the Hubert Kariuki Memorial University was accredited, as the first private University. After 2000 the political and economic wave of change catalyzed the growth of the education industry. Since then Tanzania witnessed the mushrooming of public and private universities and colleges. Within the seven years of time lapse the institutions for higher learning had added ~~to~~ more than a total of nine public universities and thirteen private universities. Also, four technical institutes and seventeen other institutions have been recorded (www.msthe.go.tz). At the same time there has been a deliberate move in several institutions to expand the student intake.

The impressive expansion of institutions for higher education had some increase of institutions that can address the development of science and technology through the teaching of microelectronics. A survey of institutions that was reported for years 2004-2005 show that out of 43 institutions of higher learning in the country about half of them are accredited to teach basic sciences and technology related subjects (www.msthe.go.tz). The figure is reduced if one is to count the number of institutions that teach microelectronics and related subjects.

Microelectronics Teaching

Microelectronics and related courses are taught at engineering and technology universities as well as at faculties of science and informatics. In Tanzania a survey shows that there are only 6 institutions that teach electronics and electronics related courses (www.msthe.go.tz). These include the University of Dar es Salaam, College of Engineering and Technology, St Joseph's University, Tanzania Open University, International Medical and Technology University and the Dar es Salaam Institute of Technology. This number of institutions may sound small but it is relatively large compared to the number of institutions that were available seven years ago. It may be noted that the contents of the course in the programs for each institution do vary in depth, ranging from diploma to postgraduate degree level material contents.

In parallel to the above mentioned training at national institutions there has been a practice of sending students to study abroad in various disciplines of national interest including the science and technology related courses. This arrangement has an added advantage to Tanzania when it comes to technological transfer.

Microelectronics Research Prospects

Students at postgraduate level were lucky to get the opportunity to work in microelectronics and material science laboratories. Thus they were exposed to the frontier knowledge of micro- and nanoelectronics and photonics and could see clearly the trend of research in the field. Few postgraduates managed to work on research areas that produced papers, hence contributed new knowledge to the international scientific community (Malisa, 2005). Some had plans to continue with such research activities at home but failed to do so due to lack of laboratory facilities to man the high technology research. Few of them managed to continue along their research fields by collaborating with their former laboratory counterparts.

Of the long list of research and teaching institutions reported only the engineering and science biased universities do research in electronics. But when it comes to micro- and nanoelectronics and photonics research a survey shows that these institutions do not have the required infrastructure for a meaningful research work. Therefore, to stimulate the interest of the existing scientists and engineers to do research in microelectronics it demands the availability of infrastructure, financial resources, the will and understanding of decision makers on the importance of electronics research field to be supported.

In comparison to many other universities in the country the University of Dar es Salaam has a high potential in establishing microelectronics research group that can develop research up to the level of micro- to nano- electronics and photonics. At this university there are two faculties that have programs that can be a catalyst to the research activities in microelectronics. The Faculty of Science has a Department of Physics with an active research group in solar energy and material science. This group, established over forty years ago, has good equipment for material science research such as sputtering and thermal evaporation systems for coating test samples with metals.

The group has test equipment like atomic force microscope that can monitor separations better than 20 nm and Fourier transform infrared (FTIR) system for studying sample structures as well as other test equipment for measurement of I-V characteristics and C-V characteristics. The research group has used the equipment for study PV systems, optical coatings for solar energy application such as optical window coatings, selective absorbing coatings, radiative cooling, thin films for photovoltaic applications and study of intrinsic performance of PV-CPC systems.

The other faculty is the Informatics and Virtual Education that has a Department of Electronics and Communication. This department teaches microelectronics, quantum electronics, optoelectronics, computer aided designs and very large scale integration. It is evident that this department is an ideal place for the microelectronics research. However, the department is new and it is currently searching for the establishment of meaningful research activities in the field of electronics and communication.

In view of the availability of the research facilities in the Department of Physics and the establishment of the Department of Electronics and Communication one can explore cooperation between the two departments to do some research work. The laboratory is not equipped with research equipment ideal for fabricating devices up to the level of micro- to nano- electronics dimensions but the available sputtering and evaporation vacuum systems can be used to fabricate and test some interesting micro devices that use organic and inorganic semiconductor materials, for start. To venture into device integration will demand additional equipment like photolithographic system, circuit design equipment and some test equipment of the IC characteristics. The availability of this equipment is a good starting point for microelectronics research. The demand for better facilities such as cleaner rooms etc is an infrastructure to follow.

The prospect of establishing and supporting a research group depends on many factors. That include:

(*a*) The availability of human resources specialized in this field. As sighted earlier, there is an astronomical increase of R&D institutions in the country; hence there is an observed increase of researchers in the field of electronics and material sciences. This opens the possibility of getting more institutions to join the University of Dar es Salaam in the research field, especially the new engineering and science institutions. There are already few academic members who have the background of microelectronics research and are familiar with microelectronics laboratories since they have worked in developed and developing countries (Samiji *et al.*, 2003). Such people are a critical mass for initiating research activities.

(*b*) Establishing research laboratories does cost a lot of money and it is common knowledge that it is difficult for many governments of the third world countries to invest so much in basic research and Tanzania is no exception. However, with time the Tanzania government has raised concern and has shown the will to promote both education and research by establishing a ministry and enacting a science and technology body. Through the Tanzania Commission for Science and Technology, that coordinates and monitors science and technology activities, the will of the government to support research activities in R&D institutions is anticipated. Also, COSTECH's declaration to support basic science research and its support of programs identified by the Centre for Science and Technology of the

Non-Aligned (NAM S&T) and Other Developing Countries is a clear evidence of future support of Tanzania's research activities in microelectronics.

(*c*) The trend observed of private investors in R&D institutions may extend their interest in supporting research activities in the institutions and expand the R&D linkage to the industries that are equally expanding in the country.

Conclusion

Tanzania supports the importance of electronics research in the country as stated by the Centre for Science and Technology of the Non-Aligned (NAM S&T) and is using the Tanzania Commission for Science and Technology to spearhead the objective through its relevant R&D institutions. However, it is noted that at present there is no active research group working on micro- and nano- electronics and photonic but there are favorable indications that Tanzania has aggressively indulged in improving education system that will bear fruits in the near future. It is envisaged that along this wave of educational growth human resources will be identified who, with the support of the government, can indulge themselves in microelectronics researches. It is also noted that with the availability of researchers who have worked in similar research laboratories in developed countries, once stimulated to continue with the research work, will be a seed for the research groups to crystallize out in numerous institutions.

References

Malisa, A., 2005. Josephson effect in MgB2/Pd/Fe/Nb Josephson junctions. Applied Physics Letters, 87 (21) pp. 213504.

Mukama, C., 2005. Benedict *et al.* Development of S&T system and Experience of Tanzania on S&T data Collection, Presented at the Regional Workshop on Science and Technology Statistics 17-22 September 2005, Etebbe, Uganda.

Samiji, M.E., Venter, A., Leitch, A.W.R, 2003.:"Thermal Stability of Pd Schottky contacts to p-type 6H-SiC" Presented at the 4th European Conference on Silicon Carbide and Related Materials (ECSCRM 2002), Linköping, Sweden (1-5 Sept 2002). Published in Materials Science Forum, 2003.

www.msthe.go.tz/statistics/index.asp

Chapter 17

Microelectronics in Turkey

Sedat Soydan
Semiconductor Technologies Research Laboratory,
National Research Institute of Electronics and Cryptology, Gebze/Kocaeli, Turkey
E-mail: sedatsoydan@uekae.tubitak.gov.tr

ABSTRACT

UEKAE–National Research Institute of Electronics and Cryptology- is one of the most important national institutes in Turkey. It is an affiliate of the Scientific and Technological Research Council of Turkey (TUBITAK). It produces and applies scientific and technological solutions with its qualified human resources and internationally accepted infrastructure in the areas of information security, communications and advanced electronics. Semiconductor Technologies Research Laboratory (YITAL) in UEKAE pioneers the developments of microelectronics area in Turkey. Today YITAL produces 1.5µm CMOS ASICs and has capability of producing 0.7µm CMOS ASICs within class-10 clean room in low volumes. 0.35 µm CMOS and 0.35 µm Si/Ge BiCMOS technology researches have been started by YITAL. On the other hand, YITAL works on some European Union projects and National Projects. One of the most significant National projects is Smart Card Design project. The European Union project named SCARD researched on Side Channel Analysis Resistant Design was supported by 6th Framework Programme on Research, Technological Development and Demonstration.

Keywords: *Microelectronics, Information security, Communications, Advanced electronics, UEKAE, Integrated circuits, MEMS, Research and Development, Turkey.*

Introduction

The electronic sector is a very promising sector to create the highest employment in the 21st century with it's already existing about US $1 trillion market volume all around the world. The Microelectronics industry brings prosperity, job opportunities and a high quality of living standards to the hosting communities. Now the Electronics

Industry is not a simple industry branch anymore, but rather turned into a basic and generic industry by itself to develop all the other industries. The concentration of high-skilled jobs and economic activity helps building strong communities.

Microelectronics is a knowledge-based high-growth sector, and industry that requires highly-skilled people. It is the driver of the information revolution and it is an enabling technology for almost every industrial sector, including computers, electronic equipment, telecommunications, financial services, transportation, entertainment, manufacturing, biotechnology and health care.

Turkey has a growing electronics industry, especially in the field of consumer electronics and telecommunication systems, and has considerable market share in consumer goods in Europe. It can be said that Turkish Electronic Industry produces mainly for exports. In 2005; 76.87 per cent (3417.3 million US $) of the total exported products went to European Union Countries. However, much of the Turkish Electronic Industry may be seen far from being at desirable dimensions when viewed with respect to world-wide figures and its share in the Turkish Economy.

The total production of electronic industry in 2005 increased from 6.8 billion US$ to 8.15 billion US $ with increase of 16.3 per cent as compared to 2004. Parallel to the increase of total production, an increase has been watched in exports by 10.3 per cent as compared to 2004 which increased from 4029 million US$ to 4445 million US $. Although, the reports of economics for electronics industry in Turkey haven't been published for 2006, it is guessed that total exports amount will be 5500 million US$ (an increase of 22.6 per cent in comparison to 2005). Also, the total production of electronics industry is guessed to be 9.5 billion US$ in 2006 (an increase of 16.5 per cent in comparison to 2005).

Microelectronics related associations can be classified into four main groups: universities, national research institutes, national and global industry companies in Turkey. Not only national universities and institutes but also industrial companies have made significant progress, recently. This attracts and motivates global companies to establish their design centers in Turkey due to the trust in educated Turkish engineers. ST Microelectronics, Maxim Integrated Products, and Hittite Microwave Corporation are some of these companies. Beside this, Turkish companies, such as Vestel and Arcelik have great progress in television and household consumer field. Because of need in this area, they established their Micro Electro Mechanical Systems (MEMS) and VLSI design centers, too. Since microelectronics fabrication industry does not exist in Turkey; in this sense, YITAL is unique in the Country because it has a critical mass of knowledge and experience on this technology. Achievements of sub half micron CMOS IC technology by improving the RTD (Research and Technology Development) capability of YITAL will encourage electronics industry to make investments in microelectronics. The new investments on microelectronics will also lead to other industries supporting microelectronics.

Smart card is one of the areas of microelectronics as a potential candidate for secure communication since magnetic cards have became subject to a growing problem of fraud. The need to secure communication in governmental and commercial associations increases the smart card demand in Turkey, as all over the world. Thus, a common mind in UEKAE comes up with the idea of designing national smart card of Turkey, in terms of operating system as well as its microchip design.

Research Activities in Distinguished Universities in Turkey

There are many universities which provide education in microelectronics, and MEMS in Turkey. In some of these universities there are design and production laboratories. Research activities and scientific projects carried out by these universities contribute to the development of microelectronics technologies. Some of the research activities in distinguished universities of Turkey are given below:

Istanbul Technical University (ITU), Istanbul

Although the microelectronics laboratory in ITU is not active today, it was the first one established in the middle of 1970s in Turkey. The staff and students, who worked in this laboratory, learned the basic concepts of wafer processing, and produced the first working integrated circuit in the lab. They played a major role in the foundation of YITAL, the unique ASIC production laboratory in Turkey, today. Presently, in the Department of EEE, courses on microelectronics technology and VLSI design for students, from undergraduate level to graduate level *i.e.* up to M.S. and Ph.D. level, are given.

These courses still create a human source for YITAL. The VLSI design lab in the department, which was founded with the aim of supporting the education about microelectronics, provides an environment to scientific researchers. Research areas of the VLSI design lab are analog, digital, and mixed-mode IC design.

Bilkent University, Ankara

There exists a Nanotechnology Research Center at Bilkent University, which is dedicated to research on theoretical and experimental nanoscience and nanotechnology with strong emphasis on education and training. This Center is an inter-disciplinary research environment, which hosts the nanotechnology related research efforts in science and engineering faculties, and serves all departments in both faculties as well as the other Turkish universities that would like to have access to the center's facilities. The center has a total of 250 m^2 of class-100 level clean room for sub-micron lithography along with general-purpose electrical and optical characterization measurements. The main research areas of the center are metamaterials, photonic crystals, MOCVD growth, fabrication and characterization of nanoelectronic and nanophotonic GaN/AlGaN devices, high performance near-infrared semiconductor photodetectors and lasers.

Sabanci University, Istanbul

Research areas of Microelectronics Engineering Department of the Sabanci University are MEMS, VLSI Systems, Photonics, R/F microwave and Acoustics. The Department envisions the gradual development of a Microelectronics Center within Sabanci University that can act both as a center-of-excellence in the academic sense concerning these research areas, and also as a pre-incubation center for possible industrial development of new products. There exists a class 1000 clean room in the university.

Middle East Technical University (METU), Ankara

Under the constitution of Department of Electrical and Electronics of METU, METUMEMS Research Group is founded with the aim of developing Micro Electro Mechanical Systems (MEMS) technology in Turkey. There is a mass microelectronics production plant (METU-MET) for Microsystems and MEMS in METU. METU-MET Facility is a microelectronics fabrication facility for 4" and 6" wafer processing. It has 1000 m^2 of class 100 and class 1000 clean room area for fabrication and 300 m^2 of class 10000 clean room area for electrical testing of IC's and active discrete components. The factory is operated by 22 technical personnel and supported by 14 researchers from Department of Electrical and Electronics Engineering. METU-MET Facilities are currently being used to develop a number of MEMS products for commercial applications, including piezoresistive pressure sensors, capacitive pressure sensors, humidity sensors, surface and bulk micro machined gyroscopes and accelerometers and also RF MEMS devices. The design team in EEE Department is working to implement various sensors using post-CMOS process, including CMOS thermopiles and uncooled infrared detector arrays.

Another research group in METU is Quantum Devices and Nanophotonics Research Group, which works on the development of modern infrared sensor array technology for high performance third generation thermal imaging systems. Current research activities of this group include the design, growth, fabrication and characterization of novel electronic and photonic devices and sensor arrays based on III-V and II-VI compound semiconductors. The experience and capabilities of the group have matured to the level of fabricating and integrating very large format mid- and long-wavelength infrared photon sensor arrays (focal plane arrays, FPAs) for thermal imaging applications covering a wide spectrum. The current work includes growth, processing and characterizing quantum well infrared photo detector (QWIP) FPAs based on alternative compound semiconductor systems, large format HgCdTe sensor arrays, quantum dot infrared photo detectors, as well as multi-color and multi-band infrared sensor arrays capable of detecting infrared radiation in more than one wavelength range.

Koc University (KU), Istanbul

Under the constitution of KU Optoelectronics Research Center there exists Optical Microsystems Research Laboratory (OML). It focuses on design, optical testing, and characterization of MOEM (micro-opto electro mechanical) and MEMS devices, Micro-scanners, Micro optics (*e.g.* diffractive elements and micro lenses), and image quality measurements in display systems. Some of the research projects in OML are given as follows:

- ✰ Modeling and characterization of nonlinear dynamic behavior of MEMS scanners
- ✰ MOEM display system development
- ✰ MOEM Infrared camera system development
- ✰ Optical signal and image processing using micro-optics and AC-coupled focal plane detector arrays

- ✰ Integrated micro-optical and MOEM system development for image acquisition and display applications
- ✰ High-resolution imaging using microlens arrays for medical imaging applications (sponsored by NEMO)
- ✰ MEMS-based Fourier Transform Spectroscopy developments, sponsored by Fraunhofer IPMS, Germany.

Bogazici University, Istanbul

In Bogazici University, there is a VLSI design laboratory (BETA) and a Micro Electro Mechanical Systems Laboratory (BUMEMS). The VLSI design laboratory was established in 1993 with the aim of conducting research in various aspects of the ever growing field of Very Large Scale Integration and establishing links between industrial companies in order to meet their ASIC demands. With the recruitment of new faculty members with backgrounds in the field, new projects were sought for and as a consequence of the received support from various sources both internal to the university and the leading industrial companies, the VLSI design laboratory facilities are now housed in BETA, which is named after the recently established Bogazici University Electronic CAD Laboratory in July 1996. Since its establishment the VLSI design group is active in the areas of computer aided design, DSP architectures, digital building blocks, and neural networks.

The aim of BUMEMS is to do research on MEMS, circuits, MEMS-circuit integration and polymer micro fabrication. The initial focus of this lab will be mainly on polymer micro fabrication and their applications to MEMS and electronic circuitry since it involves relatively cheaper and easier fabrication methods and equipments.

The Organization of UEKAE

In the information era, the security, integrity, source, and acceptability of the information have major importance. Whether it is military or civilian, secure protection and secure transfer of information has vital importance both for organizations and countries. UEAKE produces and applies scientific and technological solutions with its qualified human resources and internationally accepted infrastructure in the areas of information security, communications and advanced electronics. UEKAE has provided complete technological independence for Turkey by utilizing the TEMPEST, Acoustics, Common Criteria and Cryptographic Algorithm Design and Test Centers and Product Development, Electronic Warfare, Microelectronics and Optoelectronics departments since its foundation in 1972.

The human resources of UEKAE, which has reached to 600 employees recently, and 76 per cent of which consists of researchers, has all adopted the common ideal of "Becoming an international technology and production center that pioneers the development of new technologies in communications and advanced electronics areas". The researchers form project based teams, under the guidance of project managers, have the opportunity to exploit their creativity in total academic freedom.

UEKAE contributes to universal science by incorporating the universities into the ongoing projects and encouraging its researchers for academic career development opportunities. UEKAE has the richest library in the areas of cryptology and information security in the country as well.

UEKAE provides feasible solutions to achieve communication and security goals. Off-line and on-line crypto equipment, secure voice, fax, data communication equipment, electronic key management and distribution system, multimedia communication over HF/UHF/VHF system are some of them.

The researchers and the management both believe in modern management systems. As a result of this, the institute has been awarded the ISO 9001-2000, AQAP-2110, AQAP-2130, TS/EN ISO/IEC 17025 and BS 7799/2002 certifications, and it's EMI/EMC and the Common Criteria Test Centers have been accredited by the Turkish Accreditation Agency.

Research Departments of UEKAE are as follows:

Network Security and Common Criteria Test Laboratory

UEKAE Network Security Department is independent from security product developers and provides network security solutions instead of marketing security products. Well trained researchers can do detailed security tests on security products and solutions at the test laboratory.

Common Criteria Test Laboratory conducts security evaluations of IT products and protection profiles using test methods derived from the Common Criteria and the Common Evaluation Methodology.

Cryptographic Test and Design Laboratory

Establishing national encryption standards, designing proprietary encryption algorithms and key exchange protocols are the design activities of Cryptographic Test and Design Laboratory of UEKAE. The laboratory performs tests using the established standards and determines the security levels of the designed algorithms and protocols, and conducts cryptographic research activities.

Software Department

UEKAE develops usable software, which are open source operating system (PARDUS), smart card operating system (AKIS), electronic signature software, IP Crypto Equipment System, and ISDN Crypto Equipment System.

Acoustic Test Laboratory

The areas of the interests of the Laboratory are intelligible speech, communicability, quality assessment, speech corpuses design and development, speech and speaker recognition, language and accent identification, artificial speech synthesis, speech and video coding, noise cancellation and embedded systems hardware design.

EMC/TEMPEST Test Center

ETTM was established in 1995 as a part of a project sponsored by Turkish Government. The main task of the center is to perform EMC and TEMPEST tests of national equipment and systems. In addition to test and measurement services according to national and civil standards; training, counseling and R&D activities in EMC and TEMPEST are also performed. EMC tests in the ETTM are accredited by TURKAK (Turkish Accreditation Association).

Optoelectronic Laboratory

In order to develop new technologies in the field of optoelectronics, basic and applied researches are carried out in accredited laboratories and facilities of the Institute. Imaging technologies for forensic investigation of documents and venues, optical communications, laser technology for quality control, color measurements constitute the basic fields of study.

In each of these fields, successful projects have been carried out and many commercial instruments have been brought in since the establishment of the Optoelectronics section in 1997. Development activities are carried out for equipment realizing optical communications over free air and prototype of equipment which provides communication at 2 Mbit/s has been developed. The subject of atmospherically data transmission using laser is regarded as the future's communication system and development efforts are carried out for equipment which operates at 155 Mbit/s.

The optoelectronics technologies have a vast amount of application areas. Two of these are criminology and the textile industry. In the area of criminology, equipment for use in forensic applications is being developed. In the area of textile industry, quality control equipment which detects defects in the color, size and texture of fabric by means of optoelectronic methods are being developed.

Semiconductor Technologies Research Laboratory

In the Semiconductor Technologies Research Laboratory (YITAL) founded within the research and development activities in the field of microelectronics technologies, a totally proprietary process has been developed and a silicon integrated circuit production line has been established. Since its foundation in 1981, YITAL has been researching semiconductor technologies and developing production processes in a 450 m^2 area in TUBITAK, Gebze. At the beginning YITAL was a research laboratory within class-1000 clean room. Today YITAL has capability of 0.7μm CMOS low volume ASIC production within class-10 clean room. YITAL is the unique CMOS ASIC production foundation in Turkey. The milestones from YITAL's foundation to the present are shown in Figure 17.1.

The facilities provide as self-contained production capability from full custom design to packaged integrated circuits; including all the necessary steps needed for mask making, wafer processing and testing and encapsulating.

Recently, 0.35 μm CMOS and 0.35 μm Si/Ge technology researches, funded by the State Planning Organization (DPT), have been started by YITAL. A process run of a chip which contains various random number generator architectures continues in the process line of YITAL.

YITAL has been one of nine participants of a European Union project called SCARD. In addition to process development activities, today, YITAL works on

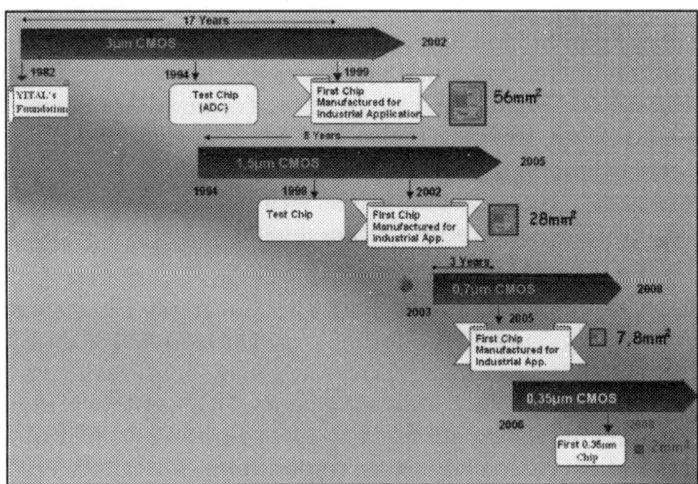

Figure 17.1: Milestones of YITAL

designing Smart Card Chip. Microprocessor, crypto co-processor, sensors, and RNG unit which are included in a Smart Card are being designed by YITAL.

Some Microelectronics Projects from UEKAE

In the following subsections, the works, which I am personally involved, carried out by UEKAE is presented in a brief format, as an example of the microelectronics activities in Turkey.

SCARD Project

Collecting information in order to retrieve secret data from a cryptographic hardware using the signals that leaks through one of its side-channels such as supply current or electromagnetic emanation is called side channel analysis (SCA). Side-channel information can be retrieved through various types of attacks (Anderson and Kuhn, 1996; Koune and Quisquater, 2002; Kommerling and Kuhn, 1999). In the invasive attacks, the attacker disassembles the device and uses expensive equipment to observe the internal nodes of the device. In the semi-invasive attacks, the attacker operates the device outside the specified ranges of operating conditions in order to introduce specific errors and this may force the device to enter an undefined state in which the device is vulnerable and leaks sensitive information. In noninvasive or passive attacks such as power analysis, the attacker tries to retrieve side-channel information using low-cost equipment when the device is working in normal conditions. This makes these attacks very popular for attackers and therefore some countermeasures should be taken at both algorithmic and physical levels of the cryptographic hardware design. Typical examples for non-invasive attacks are power attacks (Kocher *et al.*, 1999; Lash, 2002) in which the current drawn from supply pin of the device is sampled, and electromagnetic (EM) attacks in which the EM emanation of the device is observed by special probes.

A project called SCARD (Side Channel Analysis Resistant Design) was being conducted as a part of EU-FP6-STREP, with the contract number: IST-2002-507270. The consortium of the project had nine participants which are Technikon (Austria, Project Manager), IAIK (TU-Graz, Austria, Technical Manager), IFX (Infineon, Germany), TUBITAK-UEKAE (Turkey), URM1 (University of Rome, Italy), KULRD (K.U.Leuven, Belgium), UCL (U.C. de Louvain, Belgium), Cryptovision (Germany) and IICM (Austria). The goal of the project was to develop a semi-custom design flow and to design, produce and verify a secure chip through this flow using the techniques achieved in the course of the project.

At the end of the project, a chip implementing all results experienced was produced by using Infineon 0.13 μm CMOS technology. The layout of SCARD chip is given in Figure 17.2. The chip contains eight cores, seven of them are different SCA resistant implementations and the other one is a standard CMOS implementation to be reference for SCA resistant implementations. Each core comprises an 8051 μP, an AES crypto module. UEKAE-YITAL had two main responsibilities in the project. First one was to develop a secure SRAM which is resistant to side channel attacks, and the second one was analyzing some parts of the chip. The secure SRAM developed by YITAL is tied to static dual rail logic core (dr_ram) in SCARD chip.

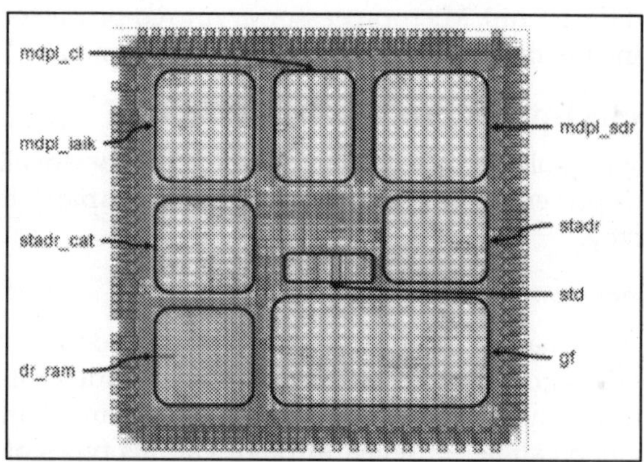

Figure 17.2: Layout of SCARD Chip

To analyze the performance of the SCA resistant implementations, a board was designed by University of Rome. They also designed an active measurement circuit, which was placed on the board. While analyzing the chip, only one core is selected via jumpers on the board. It gave the opportunity to be able to analyze the cores individually.

YITAL was responsible to analyze the Dual Rail Logic (stadr) core. We did some differential power analysis measurements on both dr_ram and std (standard CMOS for reference) cores. We analyzed the MOV operation of 8051. The oscilloscope used for measurements had sampling rate of 2 GS/s, and resolution of 8 bits. We measured the current which was drawn from VDD pin of the chip. All of the measurements were realized by programming the firmware of 8051 microprocessor.

Parallel port of the µP is used for triggering. It helped us to determine the clock cycle to be examined. We used difference of means technique in our side channel analysis. In this technique, for a reference data, a number of measurements are done in same conditions, and then the mean of the measurements is calculated. After that, the same procedure is repeated for data to be analyzed.

Finally, when the difference of means is plotted, if there is a side channel leakage for the data being processed, a peak is observed from the difference of means of power traces.

Result of difference between writing 0xFF and writing 0x00 to accumulator in standard CMOS core is given in Figure 17.3. It was observed that only with 20 measurements (10 for "mov a, #0xff" and, 10 for mov a, #0x00), standard CMOS implementation leaks information from its side channels.

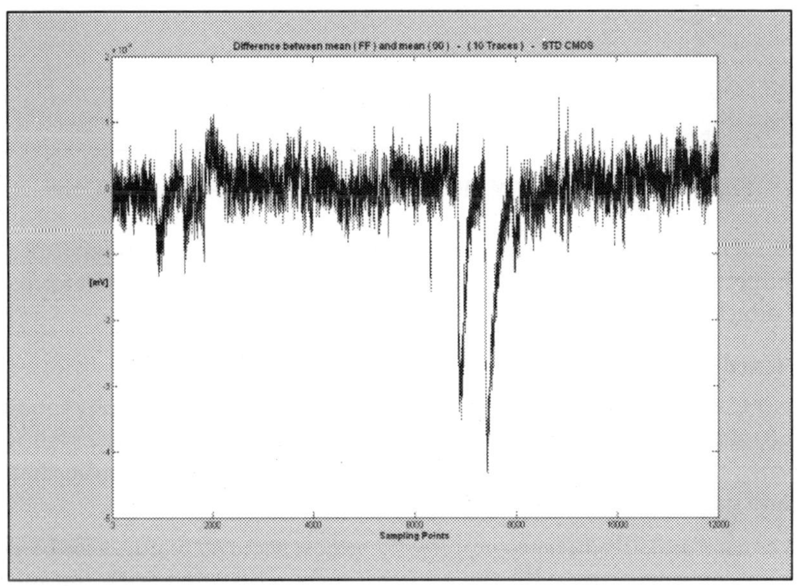

Figure 17.3: Analysis of MOV Operation on standard CMOS core

When the same measurements are repeated for Dual Rail Logic core, totally 400 power traces are needed to determine the peak. The result for DRP core is given in Figure 17.4.

As a conclusion for the differential power analysis experienced by UEKAE, the results given below were obtained:

- ☆ Dual Rail Logic core is more secure than standard CMOS core when performing mov operation. However with a sufficient number of traces Dual Rail Logic core begins to leak side channel information.
- ☆ Measured DC current for Dual Rail Logic core is 10 times higher than standard CMOS core. (0.37 mA for standard CMOS, 3.22 mA for Dual Rail Logic core)

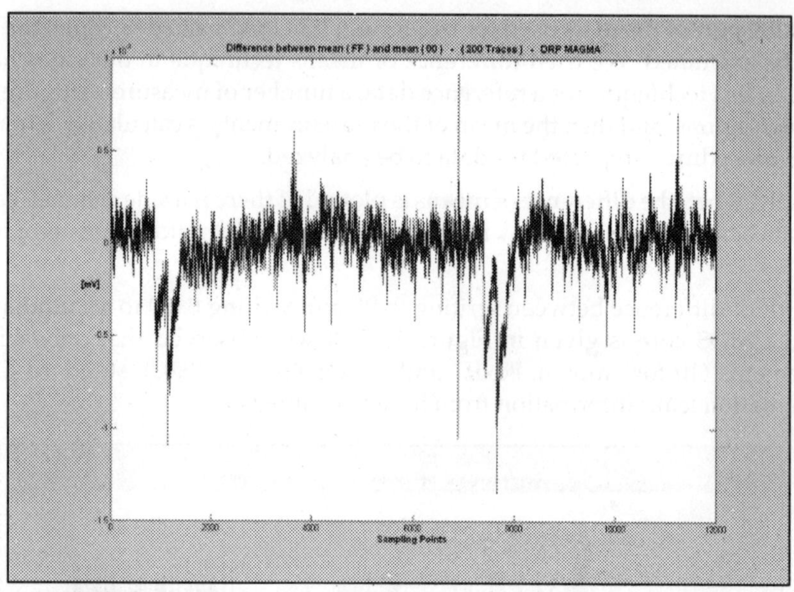

Figure 17.4: Analysis of MOV Operation on Dual-rail Logic Core

Although, the SCARD project completed in the middle of 2006, Network Security and Common Criteria group still continues to analyze SCARD chip in terms of side channel attacks to establish a new laboratory under UEKAE.

Smart Card Project

Smart Card made a high level of security (based on cryptography) available to everyone, since it can safely store secret keys and also execute cryptographic algorithms. In addition, Smart cards are so small and easy to handle that can be carried and used everywhere by everyone in everyday life.

It was a natural idea to try to employ these new security features for bank cards, in order to come to grips with the security risks associated with the increasing utilization of magnetic-strip cards.

The specific properties of smart cards, compared with all other types of cards, are determined by a microcontroller integrated in a card, which controls, initiates, and monitors all activities.

The most important part of a microcontroller is a CPU surrounded by four functional blocks: ROM, EEPROM, RAM, and an I/O port. The card also has special circuits for security and power control.

Some manufacturers offer additional functions in the form of hardware on chip. Two functions are particularly in demand: random number generators (RNGs) and cryptographic coprocessors. Some smart card microprocessors use a RNG for public-key cryptography. Countermeasures against side-channel attacks also require large quantities of readily available random numbers.

Cryptographic procedures always require random numbers. Smart cards require random numbers for:

- ☆ Key generation to authenticate the card and terminal;
- ☆ Creating padding bytes and blinding values for encryption, as initial values for transmission sequence counters; and implementation of algorithmic countermeasures against side-channel attacks.

The aim of UEKAE with the design and development of Smart Card project is covering the national needs in the following fields such as chip card applications, information security, identification, health care and transportation. This project is also funded by the State Planning Organization (DPT).

This project is still in the maturing and development period. The planned manufacturing time is at the end of 2007 and the foreseeing conclusion will be at the end of 2008.

Conclusions

In conclusion, microelectronics activities in Turkey can be summarized as the following:

- ☆ Design centers of international industry companies: Development and design of integrated circuits for global markets.
- ☆ Design centers of national industry companies: Development and design of ICs and MEMs for national and global markets.
- ☆ National research institutes: Research, development, design and fabrication of ICs.
- ☆ National and private universities: Scientific researches and applications of ICs and MEMs performed by national and international partners in VLSI and clean rooms of universities.

Even though global standards are matched in VLSI design field, in Turkey, the number of qualified engineers should be increased and industry should be encouraged for manufacturing. Both in design and manufacturing field, international corporations have significant importance.

References

Anderson, R., Kuhn, M., 1996. "Tamper Resistance – a Cautionary Note", Proceedings of the second USENIX workshop on Electronic Commerce, Oakland, California, pp. 1-11.

http://mems.ku.edu.tr/ (accessed 10 March 2007)

http://www.ee.boun.edu.tr/research.htm (accessed 10 March 2007)

http://fens.sabanciuniv.edu/micro/eng/ (accessed 12 March 2007)

http://www.nanotechnology.bilkent.edu.tr/ (accessed 12 March 2007)

http://www.ehb.itu.edu.tr/ (accessed 12 March 2007)

http://www.microsystems.metu.edu.tr/(accessed 15 March 2007)

http://www.eee.metu.edu.tr/~besikci/group/index.html (accessed 15 March 2007)

http://www.tesid.org.tr/(accessed 18 March 2007)

Koeune, F., Quisquater, J.J., 2002. "Side Channel Attacks", Scientific Report, K2Crypt.

Kommerling, O., Kuhn, M.G., 1999. "Design Principles for Tamper-Resistant Smartcard Processors", Proc. USENIX Workshop on Smartcard Technology, pp. 9-20.

Kocher, P., Jaffe, J., Jun, B., 1999. "Differential Power Analysis", Proc. of Advances in Cryptology, Lecture Notes in Computer Science, pp. 388-397.

Lash, T., 2002. "A Study of Power Analysis and the Advanced Encryption Standard", MS Scholarly Paper, George Mason University.

Chapter 18

Overview of the Status of Microelectronics, Nanoelectronics and Photonics in Uganda

Richard Tushemereirwe
Science and Technology State House, Kampala, Uganda
E-mail: richard@statehouse.go.ug, richadt2002@yahoo.com

ABSTRACT

A brief review of the global scientific and technological developments, which have led to the micro and nanoelectronics and photonic revolution, and how these are affecting the society at large. With an annual growth rate of some 15 per cent over the past three decades, the global microelectronics industry has become a central strategic enabler for the entire global economy. Public funding for R&D in micro-/nanoelectronics reaches more than €1 billion per year in the US and Japan. Asia is providing tax incentives, loans and direct support of similar orders of magnitude. New technologies are being developed for imaging, light sources, optics, and electronics. In this direction, examining the status of African science and technology, points out that it is in a dismal state in terms of the miniscule numbers of its skilled personnel in research and development, poor and neglected quality of its infrastructure and low level of student enrollment at all levels of education. Looking into Uganda's experience it is stated that there is a need for better use of knowledge, and more and better-qualified human resources for science and technology. While new and emerging technologies like nano and microelectronics, biotechnology, photonics and new materials have been receiving increasing attention in the developed world, in developing countries, there is urgent need to sensitize the masses on the importance of this field.

Keywords: *Microelectronics, Nanoelectronics, Photonics, Uganda, Science and technology, Economic development.*

Introduction

Scientific and Technological Revolution

The scientific revolution is thought to have started as early as 1543. However, historians of science disagree with this date as they see some elements contributing to the revolution as early as the 14th century. By definition, the "Scientific Revolution" refers to historical changes in thought and belief, to change in social and institutional organization, that unfolded between roughly 1550-1700; beginning with Nicholas Copernicus (1473-1543), who asserted a heliocentric (sun-centered) cosmos, it ended with Isaac Newton (1642-1727), who proposed universal laws and a Mechanical Universe (www.clas.ufl.edu).

Tomovic (1982) states that our technological era begun with the invention of the steam engine and automated regulator devices in the mid 18th century. However, other technologists believe that the first instances of technological development begun when individuals produced specialised products that other people depended upon for an improved quality of life. The development of a technological revolution is therefore a matter of degree rather than a moment in time. Early Egyptians exhibited technological shrewdness in enabling complex engineering techniques in the construction of pyramids. Technology has therefore been with humans from the first use of a stone as a tool. In a technological civilization, this tool use has developed to an advanced state, where our human culture is dependent upon the technology that surrounds it.

Human Civilization

Humanity has passed through three major evolutionary phases – hunter gatherer, agricultural and technological. For the first hundred thousand years or so of human evolution, our species existed in equilibrium with the surrounding environment as hunters and gatherers. With the introduction of agriculture people were able to settle in a particular place, as they did not have to travel in search for food. This sedimentary way of life formed the basis for modern civilization (civilization is a Latin word "civitas" meaning city). People freed from the labor of food gathering were able to pursue crafts and unique skills, so initiating the development of technologies that characterize modern life (www.ecotao.com).

Major Milestones in Human Civilization

Science and technology have profoundly influenced the course of human civilization. As early as 5000 years ago, Egypt and Mesopotamia had established irrigation systems. After the plough was invented around 5000 years ago, crop production increased dramatically (Chiras, 1994). In China, the iron plough was developed 2,600 years ago, replacing wood and stone plough as a more effective tool (Ingpen and Wilkinson, 1993). A great change followed the plough as productivity of the land increased. There have been many prehistoric inventions such as the use of fire, the use of metals such as gold and copper, bows and arrows, the fish hook, spinning and weaving, sail boats and ships, pottery, clothing, arithmetic etc.

Chemistry became an increasingly important aspect of scientific thought in the course of the sixteenth and seventeenth centuries. A number of researchers became actively engaged in chemical research. Among them were the astronomer Tycho Brahe, the chemical physician Paracelsus, and the English philosophers Robert Boyle and Isaac Newton. Mathematics was held to produce an inferior form of knowledge; mathematics could only describe, and sometimes predict, observed phenomena (Kuhn, 1970). Scientific knowledge, according to the Aristotelians, was concerned with establishing true and necessary causes of things.

New scientific developments include some of the following:–De Humani Corporis Fabrica (the Fabric of the Human Body) (1543). He found that the circulation of blood resolved from pumping of the heart. He also assembled the first human skeleton from cutting open cadavers.William Gilbert (1544-1603) Magnet and Magnetic Bodies. Galileo Galilei (1564-1642) improved the telescope and made several astonishing astronomical observations such as the phases of Venus and the moons of Jupiter. He developed the laws for falling bodies based on pioneering quantitative experiments which he analyzed mathematically. Isaac Newton (1642-1727) developed calculus and opened up new applications of the methods of mathematics to science.

The invention of the transfer resistor or "transistor" in 1947 by Bell Laboratory researchers heralded in a new era of solid-state electronics. The concept was based on the fact that it is possible to selectively control the flow of electricity through a material such as silicon, a solid material with unique conducting properties, designating some areas as conductors of current and adjacent areas as insulators thus the term "semiconductor." Recent developments include; Adrenaline (1897) John Jacob Abel, calculating machine (1614); John Napier, computer (1830); Charles Babbage, telegraph (1837); Samuel F.B. Morse, Radioactivity, Plate Techtonics (1912-1915); Alfred Wegener, Fibre optics (1955); Narinder Kampany, Electron (1897). While these developments have been highly satisfying, one is also aware of the dramatic changes that have taken place, and continue to do so, in the practice of science, in technology development, and their relationships with, and impact on, society.

Micro and Nanoelectronics and Photonic Revolution

Microelectronics

It is sixty years since the invention of the transistor at Bell Laboratories in the United States. Today there is almost no area of our lives that is not touched by microelectronics. The first computers were built in the 1940's the largest of these was called ENIAC (Electronic Numeric Integrator and Calculator). It contained 19000 vacuum tubes, would have filled a double garage and consumed enough electricity to power a small village. It broke down on average once every ten minutes and required three men working full time just to keep it operational. Today a small hand held personal organiser, can outperform this dinosaur (Halliday, 1997). What is responsible for the amazing power of today's personal computers, the humble semiconductor crystal? Microprocessors made possible the advent of the micro computer in the mid 1970s.

Over the last sixty years physicists have learned much about the fascinating properties of these crystals. The most common semiconductor is silicon, used in the majority of microelectronic devices. This has the same crystal structure as diamond (which is made from carbon atoms). The microelectronics revolution has led us into the Information Technology age where our society has developed an almost insatiable appetite for vast amounts of information.

Nanoelectronics

Nanoelectronics represent a strategic technology considering the wide range of possible applications. These include telecommunications, automotive, multimedia, consumer goods and medical systems. Nanoscience and nanotechnology are new approaches to research and development that aim to control the fundamental structure and behaviour of matter at the level of atoms and molecules. Applications of nanotechnology are emerging and will impact on the life of every citizen.

Photonics

Photonics is the use of light, rather than electricity, for transmission of data. The use of photons over electrons is what lets fiber optics handle high-bandwidth communications over long distances, but expense and complexity limit their use. Photons can travel farther than electrons (in the proper medium), use less power and generate less heat. Unlike electricity, light beams can pass through other light beams without interacting, so there's no interference. Is the fate of copper interconnecting sealed? (Krewell, 2005). The fundamental laws of physics dictate that photonic communications should eventually replace electronics, once certain obstacles are overcome. Intel is also exploring techniques to modulate optical transmissions, guide light in silicon structures and mechanically mate fiber to silicon. The photonics technology and its applications include 6 fields/categories: (1) OE components (2) FPD (3) optical input/output devices (4) optical storage (5) optical fiber communication (OFC) (6) optical components and laser applications.

World Economy and Current Trends

Micro and Nanoelectronics and Photonics in the World Economy

Technological innovations in the electronics sector have had a profound and wide-ranging impact on economic growth, foreign trade, production, skill requirements, etc. in most economies. These technologies have increased output both in terms of quality and quantity, product commercialization and trade. With an annual growth rate of some 15 per cent over the past three decades, the global microelectronics industry has become a central strategic enabler for the entire global economy, surrounding us in our daily lives as an essential constituent of a huge range of products and services: cars, phones, medical systems, multimedia applications, to name but a few since computing power is embedded in most products today. Current investment in electronics accounts for some 30 per cent of overall industrial investment in the developed world. The microelectronics value chain – from the semiconductor chip manufacturers together with their equipment and materials suppliers to the large set of related industries, such as design houses, systems builders and integrators – represent nearly 1 per cent of global gross domestic product (GDP).

In Europe alone, some 40 per cent of the annual sales of the semiconductor manufacturers are reinvested in R&D and improved production processes. As a direct result, the worldwide annual market for electronics at just under •800 billion is now bigger even than the global automotive market (Reding, 2007). When other industries that depend on electronics such as telecommunications, Internet services and consumer products to the defense and aerospace industries – are included, the global value leverages some €5,000 billion Worldwide production of transistors reached $1x10^{18}$ or a million, million, million 'gates' in 2003. By 2015, it is forecast that the industry will need capabilities to manufacture the equivalent of 10 million silicon transistors per human being per day in the developed world. For example, estimates for 2007 already indicate an increase over 2003 in demand for DVDs by 50 to 55 per cent, digital televisions by 30 to 40 per cent, mobile phones by 45 to 50 per cent and personal computers by 35 to 40 per cent . This in itself will lead to a much greater need for nanoelectronics design and manufacturing capabilities, and increased opportunities.

The US and Asia are world competitors in nanoelectronics. In 2002, the total capital spending in investments for microelectronics by the Asia/Pacific and China region amounted to 62 per cent, the US 20 per cent and Japan 10 per cent, while Europe counted for 8 per cent . Public funding for R&D in micro-/nanoelectronics reaches more than €1 billion per year in the US and Japan. Asia is providing tax incentives, loans and direct support of similar orders of magnitude. In Europe, the amount of yearly public funding has been around €600 million (Reding, 2007).

In the area of information technologies, data storage media with very high recording densities and new flexible plastic display technologies are seen as areas of major impact. In the long-term, the realization of molecular or bimolecular nanoelectronics, spintronics and quantum computing could open up new avenues beyond current computer technology.

For the optical communications industry to prosper in the long-term, electronics and photonics must converge and result in a new breed of telecom devices that are cheaper to manufacture in much larger volumes. Optoelectronics is the combination of photonics and microelectronics. When they are packaged together, they provide the capacity to generate, transport and manipulate data at phenomenal rates. Optoelectronic devices are electrical-to-optical or optical-to-electrical transducers, or instruments that use such devices in their operation. Continued development in these areas promise lower-cost optical components, such as semiconductor optical

Advanced research into manufacturing process technology is a driving force behind Europe's significant scientific and manufacturing economies. Going to smaller circuit feature sizes in the nanometer range down to 22nm or even lower, the process technology for nano-lithography, as well as for the deposition and etching of device layers, also needs to be improved. Certain of the device layers will have a thickness of only one or a few atomic layers and their deposition process needs to be very well controlled and take place in an ultra-clean environment. Obtaining the fundamental insights that will lead to acceptable manufacturing yields for the resulting billion transistor devices will be extremely demanding.

Current World Trends

There is a strong drive towards further product miniaturization. This is even leading to full integration in single 'system-on-chip' (SoC) components that can include wireless communication modules, sensors – for example: image sensors for photo and video equipment, mechanical, chemical and biosensors – and actuators such as micro-lasers, micro-switches or micro-pumps. One strong industry requirement is that the production processes be compatible with silicon wafer technologies to allow a smooth transfer in manufacturing. New technologies are being developed for imaging, light sources, optics, and electronics. New supply chain models are enabling startups to access this market as never before.

Market Analysis

The market analyst firm NanoMarkets, forecasts that the worldwide plastic electronics market will grow to US$5.8 billion (€ 4.6 billion) in 2009, and reach US$23.5 billion (€18.8 billion) by 2012. The global market for nanophotonic devices is projected to rise at an average annual growth rate of 85.8 per cent from $420.7 million in 2004 to $9.325 billion in 2009. Nanophotonic light-emitting diodes, with flat panel and plasma display applications, accounted for more than three-quarters of the market in 2003. Near-field optics (18.7 per cent) and nanocrystalline dye-sensitized solar cells (4.9 per cent) accounted for the remainder. Light-emitting diodes are not only the largest, but also the fastest-growing nanophotonic market segment, with a projected average annual growth rate of more than 90 per cent between 2004 and 2009. Near field optics and nanophotonic integrated circuits are the only other device types that are projected to have market shares greater than 1 per cent in 2009 (McWilliams, 2005). The value of the global photonics market in 2004 was US$ 203 billion (production value), a growth of 15 per cent as compared to that of 2003. In the past few years, the average growth rate of the global photonics market was about 10 per cent .

The automotive electronics market is a fast growing sector, which presents great opportunities to innovative companies that are able to meet the demanding specifications required by the industry. Market analysts estimate that by 2010 software will account for 13 per cent of the cost of a car and electronics hardware will account for 22 per cent of the total cost. Technology convergence is at a critical stage in this sector due to factors such as limited space, harsh operating environment and the need for wireless and mobile technologies. Standardization of hardware and software protocols is essential and the industry structure is changing such that traditional Original Equipment Manufacturers (for example Bosch, Siemens, Delphi, and Mercedes) need to engage with young, dynamic electronic and software companies.

Nanotechnology in Human Advancement

Technology has become so advanced that we can now stop birth defects. We can perform brain surgery. We can break free from gravity's hold and leave the atmosphere of our planet. We can change other animals (artificial selection), and control them.

We have discovered the technology and knowledge of how to clone. Genetic selection, in addition to reducing diseases through elimination of undesirable genes, will also improve physical characteristics and intelligence. In a few decades, maybe, even years, choosing the physical and mental characteristics of a child through cloning, stem cells, nanotechnology and genetic selection will be vastly preferable to the present day biological lottery. In fact, the technology has advanced to the point where adults can benefit from gene manipulation as well (Pinto, 2002).

A landslide of discoveries brought the promise of powerful electronic and computing devices, built at the molecular scale, to the forefront of scientific research in 2001 (Amarelo, 2002). These devices exist in a realm of miniaturization known as the mesoscale, between one and 1000 nanometers in size (The diameter of a human hair is 150,000 nanometers, by comparison). These computers would contain miniature chips packed with circuitry a hundred thousand times denser than today's best silicon chips.

Carbon nanotubes are some of nanoelectronics' most important construction materials, and can be used to create electronic devices and circuits at the single-molecule level. Advancing technology may end or extend life, but it can also change its quality, Products based on nanotechnology will permeate the daily lives of people who choose to use them. With humanity's newfound ability to change the world and our selves, the next century or even decade will be a new age of self-discovery. No longer will we have to adapt to what surrounds us; we could just create something to do the work for us. Technology will be the one in charge.

Africa and Other Developing Countries' Participation in Science and Technology Advancement

Africa has been grappling with the question of modernity and modernization ever since its encounter with modern Europe in the 15th century when the continent facing increasingly industrial and imperial societies was forced to reckon with its own state of economic, technological and social development (Zeleza and Kakoma, 2004). In the 19th century, the question of reforming African societies pre occupied African leaders and intellectuals from those rooted in the old scholarly traditions to intellectuals newly exposed to western education. This quest for modernity, one that was once global and distinctly African intensified during the 20th century a period that was marked by colonialism and independence. At the center of it all was the education, Science and Technology seen as vehicles for intellectual enlightenment, social engineering, cultural production, political participation and economic development.

By almost any measure, African S&T are in a dismal state; in terms of the miniscule numbers of its skilled personnel in research and development, poor and neglected quality of its infrastructure; low level of student enrollment at all levels of education; and the paltry investment in university and R&D institutes. With 13 per cent of the world's population Africa has only 0.36 per cent of the world's scientists and engineers. Africa, till now has not had a clear definition or characterization of its science and technology development. Credible indicators in science and technology are also still hard to collect (Hassan, 2002).

Nevertheless, the situation has not been and need not be as bleak. A number of organizations committees, projects, academies and institutions have been created to help promote science and technology in Africa. These include some of the following; NEPAD, UNESCO, African Academy for Science, AU, etc. More so, a number of countries have created ministries of science and technology, ICT ministries, academies of science, departments etc. For instance the New Partnership for Africa's Development (NEPAD) and related initiatives, such as the United Nations Millennium Development Goals (MDGs) and the Plan of Implementation of the World Summit on Sustainable Development (WSSD) requires renewed political and financial commitments to the development and application of S&T at national, regional and continental levels. Science and technology will play an important role in Africa's efforts to eradicate poverty, achieve food security, fight such diseases as malaria, tuberculosis and HIV/AIDS, reverse environmental degradation, and increase the pace of industrialization.

NEPAD recognizes that science and technology are central to its goals of promoting economic recovery, poverty reduction, better human health, good governance and environmental sustainability in Africa. One of its overall objectives is to bridge the technological divide between Africa and the rest of the world. It calls for the formulation and implementation of measures to: "promote cross-border co-operation and connectivity by utilizing knowledge currently available in existing centers of excellence in the continent"; and "generate a critical mass of technology expertise in targeted areas that offer high growth potential, especially in microelectronics, biotechnology and geo-science" (NEPAD, 2003).

However, there are various challenges facing the development and application of science and technology in Africa including some of the following: weak links between science institutions and private sector, relatively low or limited public and private sector expenditure on research and development (R&D), outdated science and technology policies in most countries, limited public understanding of science and technology, 'brain drain' associated with African scientists, engineers and technicians leaving the continent to work in other regions of the world; and weak and thinly spread R&D institutions.

Uganda's Experience and Future Expectations

Uganda is well placed to begin using science and technology (S&T) more intensively to achieve economic growth and social benefits. The formal sector of the economy is expanding, and real investment is going up. To sustain this growth, there is need for better use of knowledge, and more and better-qualified human resources for S&T. Currently, a few pockets of excellence stand out in a S&T system that is generally too small, under-funded, unable to keep up with knowledge advances made elsewhere in the world, and constrained by numerous disincentives and obstacles.

Uganda does not produce any sophisticated technological products due to lack of qualified personnel, lack of an Intellectual property policy to facilitate technology transfer, lack of venture capital for ST investment, lack of raw materials in the electrical sector etc. Most of the research done in the country is in the health, agriculture and education sectors. Education coverage is expanding tremendously in the country, with close to a quadrupling of enrolment at primary and tertiary levels in less than a decade. There are still many outstanding challenges that impinge on the S&T sector and these include some of the following: Very few science degree programs exist; enrolment in basic sciences is miniscule; funding for either capital or recurrent expenses for S&T training is meager; despite burgeoning enrolment, no systematic attention is being given to the development of domestic graduate education; The country has less than 500 professors; Essentially all research funding comes from external (donor) sources; Secondary-level science education is constrained by lack of laboratories and equipment, obsolete curriculum, and inadequate supply of trained science teachers.

In order to strengthen S&T systems there is need to (i) increase the amount and quality of human resources trained at undergraduate through Ph.D. level; (ii) assure quality and relevance of the research and training done; (iii) create linkages to the needs of the private sector and increase the ability of firms to use knowledge; and (iv) strengthen ties to global knowledge, through greater collaboration. Some of these objectives will be funded by the World Bank co-financed funds under the Millennium Science Initiative. The government through the Uganda National Council for Science and Technology (UNCST) has developed the National Science, Technology and Innovations Policy and ICT policies. These Policies are aimed at enhancing the science and technology system in order to improve its contribution to economic growth.

Recommendations

Science and technology have had unprecedented impact on economic growth and social development. Knowledge has become a source of economic might and power. This has led to increased restrictions on sharing of knowledge, to new norms of intellectual property rights, and to global trade and technology control regimes. Scientific and technological developments today also have deep ethical, legal and social implications. The ongoing globalization and the intensely competitive environment have a significant impact on the production and services sectors.

UNESCO's Physics Action Plan has continuously advocated for North-South partnerships with emphasis on the very difficult situation of universities in developing countries as well as to enhance communications among all physicists. An example of such collaboration is the establishment of the Edward Bouchet ICTP institute with the purpose of facilitating scientific partnerships between African and African American physicists and applied mathematicians.

Technology transfer from the North to the South is an important issue in the Asian context and may become more important if other developing countries resume their growth process and specially if the industrialization process is redirected towards the creation of dynamic comparative advantages. While new and emerging technologies like nano and microelectronics, biotechnology, photonics and new materials have been receiving increasing attention in the developed world, in developing countries, there is need to sensitize the masses on the importance of this knowledge.

Developing countries should invest atleast 1 per cent of their GDP in to R&D in order to survive in this information age. India has invested atleast 0.75 per cent of its GDP in research and today, India is one of the leading countries in information communication technology. Developing countries should provide a policy environment that is conducive to scientists and investors in science and technology.

Conclusions

We know that, if correctly employed, science and technology can lead to the betterment of the human race, to the development of the qualities of humanity, and to an understanding of the mysteries of the universe. We know that it has the potential to eradicate poverty, enrich humanity, and free it from the struggle for existence. It is through science that we can understand the potential of existing resources and learn to develop this natural heritage for ourselves and future generations. Science should, therefore, be pursued to improve human life, and have as its conscious and ultimate goal the establishment of world peace and the unification of the human race.

References

Amarelo, Monica, 2002. American Association of the advancement of science.

Chiras, D.D. 1994. Environnemental Science. Action for a sustainable future. The Benjamin/Cummings Publ. Co. Ltd

Chyba, C. F., Thomas, P. J., Brookshaw, L., Sagan, C., 27 July 1990. Cometary delivery of organic molecules to the early Earth. Science. Vol.249, no.4967. pp. 366-73.

Hunter, L. 1997. The Cheetah. Racing towards extinction or adaptable specialist? In: Africa. Environment and Wildlife. V5 (6).

Halliday, Douglas, 1997. The microelectronics Revolution.

Hassan, Mohamed, 2002. Science and Africa's salvation.

Ingpen, R., Wilkinson, P., 1993. Encyclopedia of Ideas that changed the world. The greatest discoveries and inventions of human history. Viking Studio Books.

Kuhn, 1970. The structure of scientific revolution.

Krewell, Kevin, 2005. Is the fate of copper interconnect sealed?

McWilliams, Andrew, 2005. Nanotechnology for Photonics. Bcc. Research.

NEPAD, 2003. Ministerial conference on Science and Technology. Johannesburg, South Africa.

Pinto, Jim, 2002. Genetic competition: Technology speeds evolution.

Rainer Waser Nano, 2005. Electronics and information technology.

Reding, Viviane, 2007. Think big about small things. Europe's ambitious strategy to be world leader in nanoelectronics.

Tomovic, Rajko, 1982. Science and Technology in the transformation of the world. UNU Press.

www.rand.org/pubs/monograph_reports/The global technology revolution

www.ecotao.com/holism/hu_mod.htm. Human civilization and Modern culture, civilizations

www.indianchild.com. Famous inventions

www.nanoregnews.com Nano and Giga challenges in electronics and photonics

www.clas.ufl.edu/users

Zeleza, Paul, Tiyambe, Kakoma, Ibulaimu, 2004. Science and technology in Africa. Africa World Press.